O9-BTI-686

ENGINEERING GRAPHICS

Tools for the Mind

Bryan Graham
University of Alabama

SDC
Publications

SDC Publications
P.O. Box 1334
Mission, KS 66222
913-262-2664
www.SDCpublications.com
Publisher: Stephen Schroff

Engineering Graphics: Tools for the Mind
Copyright 2016 Bryan Graham

Technical Graphics
Copyright 2007 Meyers, Croft, Miller, Demel & Enders

All rights reserved. This document may not be copied, photocopied, reproduced, transmitted, or translated in any form or for any purpose without the express written consent of the publisher, SDC Publications.

It is a violation of United States copyright laws to make copies in any form or media of the contents of this book for commercial or educational purposes without written permission.

Examination Copies:
Books received as examination copies are for review purposes only and may not be made available for student use. Resale of examination copies is prohibited.

Electronic Files:
Any electronic files associated with this book are licensed to the original user only. These files may not be transferred to any other party.

ISBN-13: 978-1-63057-086-6
ISBN-10: 1-63057-086-9

Printed and bound in the United States of America.

ENGINEERING GRAPHICS:
Tools for the Mind
Bryan Graham

Table of Contents

1. Lettering
Guidelines for
Lettering

2. Sketching
Lines
Circles
Arcs
Isometric Sketching

3. Orthographic Projection
Planes and Surfaces
Curved Surfaces

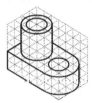

4. Isometric Drawings
Inclined Surface
Oblique Surface
Curved Surfaces

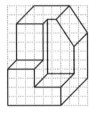

5. Oblique Drawings
Cavalier Oblique
Cabinet Oblique
Curved Surfaces

6. Auxiliary Views
Inclined Surface

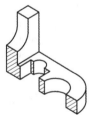

7. Sections
Full Sections
Half Sections
Offset Sections
Revolved/Removed
  Sections
Broken-Out Sections
Aligned Sections

8. Dimensioning
Basic Rules
Size
- Arc Dimension
  Placement
- Holes and Cylindrical
  Features
- Machined Holes
Locate
- Finished Surfaces
- Holes or Cylinders on
  an arc or circle
Overall Dimensions
- Rounded End Objects

Exclusive Bonus Content

Technical Graphics
Meyers • Croft • Miller • Demel • Enders
Ohio State University

Table of Contents

CHAPTER 1 Getting Started

1.1	**Introduction**	**1-1**
1.2	**What You Will Learn From This Text**	**1-3**
1.2.1	Lettering	1-3
1.2.2	Sketching Techniques	1-3
1.2.3	Shape Description	1-3
	Orthographic Projection	1-3
	Pictorial Drawings	1-4
	Perspective Drawings	1-4
1.2.4	Sections and Conventions	1-4
1.2.5	Size Description	1-4
1.2.6	Fastening and Joining	1-5
1.2.7	Drawing Notes	1-5
1.2.8	Production Drawings	1-5
1.2.9	Descriptive Geometry	1-5
1.2.10	Presentation Graphics	1-7
1.2.11	Computer Systems	1-9
	Computer-aided Manufacturing	1-10
	Computer-integrated Manufacturing	1-11
	Flexible Manufacturing Systems	1-11
1.3	**Tools You Will Need**	**1-12**
1.3.1	Tools for Sketching	1-12
1.3.2	Tools for Instrument Drawing	1-13
1.3.3	Drawing by Computer	1-15
	Central Processing Unit	1-15
	Input Devices	1-15
	Output Devices	1-16
	Data Storage Devices	1-16
	Software	1-16
1.4	**Using Your Drawing Tools**	**1-17**
1.4.1	Sketching	1-17
1.4.2	Alphabet of Lines	1-17
1.4.3	Instrument Drawing	1-17
	Straight Lines	1-17
	Drawing Straight Lines with Your CAD System	1-19
	Circles and Arcs	1-20
	Drawing Circles and Arcs With Your CAD System	1-20
	Measuring	1-21
	Measuring With Your CAD System	1-22

1.5 Lettering 1-22

    1.5.1   Lettering Guidelines 1-23

    1.5.2   Lettering Styles 1-23

    1.5.3   Character Uniformity 1-24

    1.5.4   Spacing 1-24

1.6   **Summary** 1-25

CHAPTER 2 Technical Sketching

2.1   **Introduction** 2-1

2.2   **Sketching Horizontal Lines** 2-2

2.3   **Sketching Vertical Lines** 2-2

2.4   **Sketching Angular Lines** 2-2

2.5   **Sketching Curved Lines** 2-2

2.6   **Drawing Media** 2-3

2.7   **Grids for Sketching** 2-3

2.8   **Line Weights** 2-4

2.9   **Sketching Instruments** 2-4

2.10   **Enlarging and Reducing Using a Grid** 2-4

2.11   **Enlarging and Reducing Using a Rectangular Diagonal** 2-5

2.12   **Enlarging and Reducing by Changing the Focal Distance from the Object to the Plane** 2-5

2.13   **Sketching Small Objects with a Hand for Proportion** 2-6

2.14   **Sketching Solids** 2-6

2.15   **Four Steps to Sketching an Object** 2-6

2.16   **Basic Shading of Solids** 2-8

2.17   **Orthographic Projection Sketch** 2-8

2.18   **Isometric Pictorial Sketch** 2-9

2.19   **Oblique Pictorial Sketch** 2-9

2.20   **Where Do Ideas Come From That Require Sketching?** 2-10

2.21   **Sample Engineering Applications** 2-11

2.22   **Sample Problem: Table Saw Fixture** 2-12

2.23   **Summary** 2-13

CHAPTER 3 Orthographic Projection

3.1   **Introduction** 3-1

    3.1.1   Communicating in Three Dimensions 3-1

3.2   **Theory of Orthographic Projection** 3-2

    3.2.1   First-Angle and Third-Angle Projection 3-3

3.3   **Principal Views and Planes** 3-4

    3.3.1   Front View and the Frontal Plane 3-5

    3.3.2   Top View and the Horizontal Plane 3-5

3.3.3   Side View and the Profile Plane 3-5
3.3.4   Other Principal Views and Planes 3-5
3.3.5   Relation of the Principal Views 3-6

3.4   **Principal Dimensions** 3-7

3.5   **Nonlinear Features** 3-7
3.5.1   Center Lines 3-8

3.6   **Visibility** 3-8
3.6.1   Describing Hidden Features 3-8

3.7   **The Alphabet of Lines** 3-9
3.7.1   Visible Lines 3-9
3.7.2   Dashed Lines for Hidden Features 3-9
3.7.3   Junctions of Lines 3-10
3.7.4   Center Lines 3-10
3.7.5   Construction Lines 3-10
3.7.6   Precedence of Lines 3-11

3.8   **Using Subscripts** 3-11

3.9   **What a Line Represents** 3-12
3.9.1   A Line as the Edge View of a Surface 3-12
3.9.2   A Line as an Intersection 3-12
3.9.3   A Line as a Limiting Element 3-12
3.9.4   What Can a Line Represent? 3-12

3.10   **What a Point Represents** 3-13
3.10.1   A Point as the Intersection of Lines 3-13
3.10.2   What Can a Point Represent? 3-13

3.11   **Reading the Surfaces of Objects** 3-14
3.11.1   Surfaces That Are True Size and Shape 3-14
3.11.2   Surfaces That Are Inclined 3-14
3.11.3   Surfaces That Are Oblique 3-15
3.11.4   Reading by Sketching 3-15
3.11.5   Curved and Tangent Surfaces 3-16
3.11.6   Conventional Practices 3-16
3.11.7   Fillets, Rounds, and Runouts 3-17

3.12   **Keys to Reading Three-Dimensional Objects** 3-18
3.12.1   Identify and Locate Points on the Object 3-18
3.12.2   Identify and Locate Lines on the Object 3-19
3.12.3   Identify and Locate Surfaces on the Object 3-19
3.12.4   Curved Surfaces and Limiting Elements 3-20
3.12.5   Reading an Object by Sketching It 3-20
3.12.6   Reading by Modeling the Object 3-20

3.13   **Creating a Multiview Orthographic Drawing** 3-21
3.13.1   Choice of Views 3-22
3.13.2   Planning a Drawing 3-22
3.13.3   Developing the Drawing 3-23
3.13.4   Checking Your Work 3-24

3.14   **Creating a Drawing on the Computer** 3-24
3.14.1   Selecting a Grid 3-24
3.14.2   Selecting the Layers 3-25
3.14.3   Selecting Pen Widths and Colors for a Printer 3-25
3.14.4   Developing the Drawing on the Computer 3-25
             Construction Lines 3-25
             Visible Lines 3-26

Finishing the Drawing 3-26
Alternate Methods for Layering 3-26
3.14.5 Modeling With CAD 3-28

3.15 **Auxiliary Views** 3-29
3.15.1 Preparing an Auxiliary View Drawing 3-30
3.15.2 Other Auxiliary Views 3-32

3.16 **Sample Problem: Table Saw Fixture** 3-32

3.17 **Summary** 3-33

CHAPTER 4 Pictorial Drawings

4.1 **Introduction** 4-1

4.2 **Axonometric Drawings** 4-1
4.2.1 Types of Axonometric Drawings 4-1
4.2.2 The Isometric Axes 4-2
4.2.3 Normal Surfaces in Isometric 4-2
4.2.4 Inclined Surfaces in Isometric 4-3
4.2.5 Oblique Surfaces in Isometric 4-4
4.2.6 Circles and Arcs Using an Isometric Template 4-4
4.2.7 Sketching and Drawing Circles and Arcs 4-6
4.2.8 Sketching or Drawing Ellipses on Non-isometric Surfaces ... 4-8
4.2.9 Sketching an Irregular Ellipse or Curve 4-9
4.2.10 Exploded Views 4-9
4.2.11 Isometric Assembly Sketches and Drawings 4-10
4.2.12 Isometric Projections 4-10

4.3 **Oblique Drawings** 4-12
4.3.1 Oblique Views 4-12
4.3.2 Cabinet and Cavalier Oblique Drawings 4-13

4.4 **Perspective Drawings** 4-14
4.4.1 Perspective Overview 4-14
4.4.2 One-Point Perspective 4-14
4.4.3 Two-Point Perspective 4-16
4.4.4 Layout of the Object and Picture Plane 4-17
4.4.5 The Measuring Line 4-18
4.4.6 The Horizon 4-19
4.4.7 The Station Point 4-20
4.4.8 Vanishing Points 4-20
4.4.9 Projectors 4-20
4.4.10 Visual Rays 4-21
4.4.11 Presentation Variables 4-22
4.4.12 Angular Surfaces 4-22
4.4.13 Curved Surfaces 4-23
4.4.14 Three-Point Perspective 4-25

4.5 **Computer Applications** 4-25

4.6 **Sample Problem: Table Saw Fixture** 4-25

4.7 **Summary** 4-25

CHAPTER 5 Sections and Conventions

5.1 **Introduction** 5-1

5.2 **Cutting Plane** 5-1

5.3   **Section Lines** 5-3

5.4   **Full Section** 5-5

5.5   **Half Sections** 5-6

5.6   **Offset Sections** 5-7

5.7   **Revolved Sections** 5-7

5.8   **Removed Sections** 5-7

5.9   **Broken-out Sections** 5-8

5.10   **Sectioned Assemblies** 5-8

5.11   **Sectioned Isometric Drawings** 5-8

5.12   **Conventional Breaks** 5-9

5.13   **Conventional Revolutions** 5-9
            5.13.1   Conventional Hole Revolutions 5-10
            5.13.2   Conventional Rib Revolutions 5-10
            5.13.3   Conventional Spoke Revolutions 5-11
            5.13.4   Conventional Lug Revolutions 5-12
            5.13.5   Hole, Rib, Spoke, and Lug Combination 5-12
            5.13.6   Alternate Crosshatching 5-13

5.14   **Sample Problem: Table Saw Fixture** 5-13

5.15   **Summary** 5-13

CHAPTER 6 Dimensions and Tolerances

6.1   **Introduction** 6-1
            6.1.1   Dimensioning Standards 6-1
            6.1.2   Dimensioning Practices 6-2

6.2   **Scaling and Dimensions** 6-2
            6.2.1   Types of Scales 6-3
            6.2.2   Specifying Scales on a Drawing 6-5

6.3   **Dimensioning Basic Shapes** 6-5
            6.3.1   Assumptions 6-5
                      Perpendicularity 6-5
                      Symmetry 6-5
            6.3.2   Simple Shapes 6-5
                      Rectangular Prisms 6-5
                      Cylinders 6-5
                      Cone and Frustum 6-6
            6.3.3   Simple Curved Shapes 6-6
            6.3.4   Circular Center Lines 6-7

6.4   **Placement of Dimensions** 6-8
            6.4.1   Characteristic View 6-8
            6.4.2   Dimensions Should Be Grouped 6-8
            6.4.3   Preferred Locations 6-8

6.5   **Dimensioning Systems** 6-8
            6.5.1   Alignment of Dimension Numbers 6-8
            6.5.2   Units of Measurement 6-9

6.6   **Standard Practices** 6-9
            6.6.1   Lines for Dimensioning 6-9

6.6.2   Arrowheads 6-9
6.6.3   Lettering and Notes 6-10
6.6.4   Readability 6-10

6.7   **Planning and Layout** 6-11

6.8   **Functional Relationships** 6-11
    6.8.1   Tolerance Accumulation 6-11
    6.8.2   Dimensioning to Finished Surfaces 6-12
    6.8.3   Dimensions for Locating Features on an Object 6-13
    6.8.4   Procedure for Functional Dimensioning 6-14

6.9   **Dimensioning on the Computer** 6-14
    6.9.1   A Layer for Dimensions 6-14
    6.9.2   Scaling a CAD Drawing 6-14
    6.9.3   Automatic Dimensioning 6-15

6.10   **Other Standard Dimensioning Systems** 6-15
    6.10.1   Datum Lines and Surfaces 6-16
    6.10.2   Tabular Dimensioning 6-16
    6.10.3   Other Design Disciplines 6-17

6.11   **Dimensioning for Interchangeable Parts** 6-18
    6.11.1 Types of Fits 6-18
        Clearance Fits 6-19
        Checking Fit Calculations 6-19
        Interference Fits 6-20
        Transition Fits 6-20
    6.11.2   Standard Fits 6-20
    6.11.3   Cylindrical Fits 6-21
        Basic Hole System 6-21
        Basic Shaft System 6-22

6.12   **Advanced Dimensioning Topics** 6-23
    6.12.1   Metric Dimensioning With International Standards 6-23
    6.12.2   Surface Control 6-23
    6.12.3   Geometric Tolerancing 6-23

6.13   **Principles of Good Dimensioning** 6-23

6.14   **Checking a Dimensioned Drawing** 6-28

6.15   **Sample Problem: Table Saw Fixture** 6-28

6.16   **Summary** 6-29

CHAPTER 7 Dimensioning for Production

7.1   **Introduction** 7-1

7.2   **Manufacturing Processes** 7-2
    7.2.1   Material Reformation 7-2
    7.2.2   Material Removal 7-3
    7.2.3   Material Addition 7-5
    7.2.4   Assembly Processes 7-6

7.3   **Standard Tables for Fits** 7-7
    7.3.1   Preferred Limits and Fits in Inches 7-7
    7.3.2   Preferred Metric Limits and Fits 7-9

7.4   **Surface Control** 7-11

7.5   **Geometric Tolerancing** 7-14

7.5.1   Definitions and Symbols 7-15
7.5.2   Datum Referencing 7-16
7.5.3   Tolerances of Location 7-17
7.5.4   Tolerances of Form 7-19

7.6   **Sample Problem: Table Saw Fixture** 7-22

7.7   **Summary** 7-22

CHAPTER 8 Fastening, Joining, and Standard Parts

8.1   **Introduction** 8-1

8.2   **Screw-Thread Terms** 8-2

8.3   **Other Properties of Threads** 8-3
8.3.1   Right-hand and Left-hand Threads 8-3
8.3.2   Screw-Thread Forms 8-3
8.3.3   Multiple Threads 8-4
8.3.4   Thread Series 8-5
8.3.5   Specifying a Thread 8-6

8.4   **Threaded Fasteners** 8-7
8.4.1   Screws 8-7
8.4.2   Bolts, Nuts, Studs, and Washers 8-9
8.4.3   Materials for Fasteners 8-11

8.5   **Representing Threads and Fasteners on Drawings** 8-11
8.5.1   Detailed Thread Symbols 8-12
8.5.2   Schematic Thread Symbols 8-13
8.5.3   Simplified Thread Symbols 8-14
8.5.4   Drawing Screw Heads and Nuts 8-14
8.5.5   Specifying Fasteners 8-15

8.6   **Other Mechanical Fasteners** 8-15
8.6.1   Rivets 8-16
8.6.2   Pins 8-16
8.6.3   Keys 8-17
8.6.4   Retaining Rings 8-18

8.7   **Other Machine Elements** 8-18

8.8   **Thermal Processes** 8-20
8.8.1   Welding Processes 8-20
8.8.2   Specifying Welds on Drawings 8-22
8.8.3   Soldering and Brazing 8-23

8.9   **Adhesives** 8-24
8.9.1   Principal Types of Adhesives 8-24

8.10   **The Future of Fastening and Joining** 8-25

8.11   **Sample Problem: Table Saw Fixture** 8-26

8.12   **Summary** 8-26

CHAPTER 9 Production Drawings

9.1   **Introduction** 9-1

9.2   **Applications of Production Drawings** 9-12

9.3   **Types of Production Drawings** 9-12

9.3.1 Orthographic Detail Drawings 9-12

9.3.2 Sectioned Assembly Drawings 9-13

9.3.3 Pictorial Drawings and Diagrams 9-14

9.3.4 Drawings for Other Disciplines 9-16

9.4 **Callouts** 9-19

9.5 **Identification of Measuring Units** 9-19

9.6 **Paper Sizes for Drawings** 9-20

9.7 **Title Blocks** 9-21

9.8 **Revisions** 9-24

9.9 **Parts List** 9-24

9.10 **General Notes** 9-25

9.11 **Applications of the CAD Database** 9-26

9.12 **Flow of Information** 9-29

9.13 **Accompanying Written Specifications** 9-30

9.14 **Sample Problem: Table Saw Fixture** 9-30

9.15 **Summary** 9-31

CHAPTER 10 Three-Dimensional Geometry Concepts

10.1 **Introduction** 10-1

10.2 **Applications of Descriptive Geometry** 10-8

10.2.1 Problem-solving Skills 10-9

10.2.2 Developing Visualization Skills 10-10

10.3 **Operations** 10-10

10.3.1 The Direct Method 10-10

10.3.2 The Glass Box: Fold Line Method 10-13

10.4 **Line Relationships** 10-15

10.4.1 Principal Lines 10-15

10.4.2 Inclined Lines 10-15

10.4.3 Oblique Lines 10-15

10.4.4 Parallel Lines 10-18

10.4.5 Perpendicular Lines 10-19

10.4.6 Angular Relationships 10-20

10.4.7 Applications 10-21

10.5 **Plane and Line Relationships** 10-22

10.5.1 Principal Planes 10-22

10.5.2 Inclined Planes 10-22

10.5.3 Oblique Planes 10-27

10.5.4 Lines Parallel to Planes 10-29

10.5.5 Lines Perpendicular to Planes 10-30

10.5.6 Angles between Lines and Planes 10-30

10.5.7 Lines Intersecting Planes 10-33

10.5.8 Planes Parallel to Planes 10-34

10.5.9 Dihedral Angles 10-34

10.5.10 Planar Intersections 10-35

10.5.11 Plane-Solid Intersections 10-36

10.5.12 Solid-Solid Intersections 10-39

10.6   **Curved Surfaces** **10-40**
   10.6.1   Line-Cylinder Intersections **10-43**
   10.6.2   Line-Cone Intersections **10-44**
   10.6.3   Line-Sphere Intersections **10-45**
   10.6.4   Plane-Cylinder Intersections **10-45**
   10.6.5   Plane-Cone Intersections **10-46**
   10.6.6   Cylinder-Cylinder Intersections **10-47**
   10.6.7   Cylinder-Right Circular Cone Intersection **10-48**

10.7   **Sample Problem: Table Saw Fixture** **10-49**

10.8   **Summary** **10-50**

CHAPTER 11 3-D Geometry Applications

11.1   **Introduction** **11-1**

11.2   **Pattern Developments** **11-1**
   11.2.1   Principles and Examples **11-1**
   11.2.2   Right Solids **11-2**
   11.2.3   Cylinders **11-6**
   11.2.4   Cones **11-10**
   11.2.5   Pyramids and Tetrahedrons **11-11**
   11.2.6   Transition Pieces **11-12**
   11.2.7   Models From Intersecting Solids **11-13**
   11.2.8   Applications **11-16**

11.3   **Maps and Contour Plots** **11-16**
   11.3.1   Map Introduction **11-17**
   11.3.2   Concept of Representing Elevation **11-22**
   11.3.3   Sections or Profiles and Applications **11-22**

11.4   **Vector Graphics** **11-23**
   11.4.1   Vector Addition and Subtraction **11-24**
   11.4.2   Applications **11-25**
          Statics: Two-Dimensional, Single-Joint Problem **11-25**
          Statics: Two-Dimensional Truss Analysis **11-27**
          Statics: Three-Dimensional Analysis **11-27**

11.5   **Shades and Shadows** **11-28**
   11.5.1   Shade, Shadows, and the Standard Light Ray **11-28**
   11.5.2   Orthographic Views **11-29**
   11.5.3   Shades and Shadows in Isometric Pictorials **11-37**
   11.5.4   Shades and Shadows in Perspective Pictorials **11-40**

11.6   **Sample Problem: Table Saw Fixture** **11-42**

11.7   **Summary** **11-42**

CHAPTER 12 Graphical Presentation of Data

12.1   **Introduction** **12-1**

12.2   **Charts and Graphs** **12-1**
   12.2.1   Classes of Charts and Graphs **12-1**
   12.2.2   Typical Uses of Charts and Graphs **12-2**
   12.2.3   Types of Graphs and Their Uses **12-4**
          Line Graphs **12-4**
          Multiple-Line Graphs **12-5**
          Multiple-Scale Graphs **12-5**

Pie Charts 12-6

Polar Charts 12-7

Bar Charts 12-8

Surface Charts 12-9

Three-Coordinate Charts 12-9

Tri-linear Charts 12-10

12.2.4 Plotting a Graph 12-11

Selecting the Type of Graph 12-11

Selecting a Size for a Graph 12-11

Selecting and Calibrating a Scale 12-12

12.2.5 Adding Titles and Labels 12-14

Chart/Graph Title 12-14

Labeling the Axes 12-15

12.2.6 Modeling Empirical Data (Data Scatter) 12-16

Best-Fit Methods 12-16

Types of Data 12-17

12.2.7 Computer-based Chart and Graph Programs 12-17

12.3   Merging Text and Graphics by Computer 12-20

12.4   Sample Problem: Table Saw Fixture 12-23

12.5   Summary 12-23

CHAPTER 13 Design Process

13.1   Introduction 13-1

13.1.1 Product Design and Systems Design 13-2

13.2   Scientific Method 13-4

13.2.1 Determine a Need 13-5

13.2.2 Establish Known Information 13-6

13.2.3 Stimulate Design Concepts 13-6

13.2.4 Investigate Design Concepts With Analytical Tools 13-7

13.2.5 Gather and Review Investigation Data 13-8

13.2.6 Now Test the Design 13-8

13.2.7 Propose to Accept, Modify, or Reject the Design 13-9

13.2.8 Render the Final Decision and Prepare Working Drawings 13-10

13.2.9 Optimize Design by Iteration 13-10

13.3   Lateral Thinking 13-11

13.4   Presentation 13-12

13.4.1 What Is Being Presented? 13-12

13.4.2 Why Is the Presentation Being Made? 13-12

13.4.3 Who Is the Audience? 13-13

13.4.4 When Will the Presentation Be Made? 13-13

13.4.5 Where Will the Presentation Be Made? 13-13

13.4.6 How Will the Information Be Presented? 13-14

13.4.7 How Much Will the Presentation Cost? 13-14

13.4.8 Notes on Content of Presentations 13-14

13.5   Protection of Designs 13-15

13.6   A Design Problem 13-16

13.6.1 Determine a Need 13-16

13.6.2 Establish Known Information 13-16

13.6.3 Stimulate Design Concepts 13-18

13.6.4 Investigate Design Concepts With Analytical Tools 13-20

13.6.5 Gather and Review Investigation Data 13-22

13.6.6   Now Test the Design 13-22

13.6.7   Propose to Accept, Modify, or Reject the Design 13-23

13.6.8   Render the Final Decision 13-24

13.6.9   Optimize Design by Iteration 13-25

13.7   **Summary** 13-26

APPENDIX Tables

Table A.1   **Preferred Limits and Fits (Inches)** A-3

Table A.2   **Preferred Limits and Fits (Metric)** A-10

Table A.3   **American National Standard Unified Inch Screw Threads** A-14

Table A.4   **Standard Coarse Pitch Metric Threads** A-15

Table A.5   **Standard Fine Pitch Metric Threads** A-15

Table A.6   **Dimensions of Hex Bolts** A-16

Table A.7   **Dimensions of Square Bolts** A-17

Table A.8   **Dimensions of Hex Nuts and Hex Jam Nuts** A-18

Table A.9   **Dimensions of Square Nuts** A-19

Table A.10   **Dimensions of Hex Cap Screws** A-20

Table A.11   **Dimensions of Slotted Flat Countersunk Head Cap Screws** A-21

Table A.12   **Dimensions of Slotted Round Head Cap Screws** A-22

Table A.13   **Dimensions of Slotted Fillister Head Cap Screws** A-23

Table A.14   **Dimensions of Hex Head Machine Screws** A-24

Table A.15   **Dimensions of Slotted Flat Countersunk Head Machine Screws** A-25

Table A.16   **Dimensions of Slotted Fillister Head Machine Screws** A-26

Table A.17   **Dimensions of Square and Hex Machine Screw Nuts** A-27

Table A.18   **Dimensions of Slotted Headless Set Screws** A-28

Table A.19   **Dimensions of Square Head Set Screws** A-29

Table A.20   **Shoulder Screws** A-30

Table A.21   **Dimensions of Preferred Sizes of Type A Plain Washers** A-31

Table A.22   **Dimensions of Woodruff Keys** A-32

Table A.23   **Tap Drill Sizes – Inches** A-33

Table A.24   **Tap Drill Sizes – Metric** A-33

Table A.25   **Standard Welding Symbols** A-34

Table A.26   **Fraction – Decimal Conversion Table** A-36

ENGINEERING GRAPHICS:
Tools for the Mind

LETTERING

The text for engineering drawings is made with single strokes and is known as "Gothic" lettering. The purpose of the simple, sans serif lettering is to provide easily read and understood information to supplement the drawing and indicate dimensions for the manufacture or fabrication of the drawing.

Figure LTR-2 and **Figure LTR-3** illustrate the order of the strokes for each letter and numeral in order to practice lettering. Notice that the letters are typically made with more than one stroke of the pencil. Large letters are used for practice to learn the technique of constructing each letter. After your lettering becomes consistent and uniform, practice with smaller letters representative of what you would find on an engineering drawing.

Notes and dimensions are typically 3 mm or 1/8" inch in height. For reference, the grids in this workbook are typically spaced 1/8" inch (or 3 mm) apart.

As you look at the illustrations in the workbook, notice the spacing between letters is consistent. However, for some letters, like a "V" placed next to an "A", the spacing is adjusted to avoid excessive negative space between the letters. This concept is called kerning **(Figure LTR-1)**.

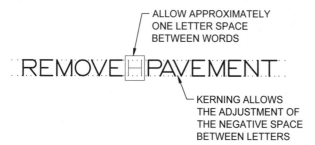

ALLOW APPROXIMATELY ONE LETTER SPACE BETWEEN WORDS

KERNING ALLOWS THE ADJUSTMENT OF THE NEGATIVE SPACE BETWEEN LETTERS

Figure LTR-1 — Kearning involves the proportional spacing of letters.

Allow extra space between words. This extra space is approximately equivalent to the width of a full letter (like the letter "H") with the additional spacing on each side of the letter. The end of a sentence and the beginning of the next sentence are separated by approximately two letter spaces.

Letter placement becomes faster and more uniform as lettering is practiced and used.

GUIDELINES FOR LETTERING:

1. Use even pressure to draw precise, clean lines.

2. Keep the pencil sharp for clean, legible lines.

3. Use one stroke per line.

4. Horizontal strokes are drawn left to right.

5. Vertical strokes are drawn downward.

6. Curved strokes are drawn top to bottom with one continuous stroke on each side.

7. Spacing between letters is based on letter width, with the letter "W" being the widest and the letter "I" being the narrowest.

8. Spacing between words should be the width of one letter with space n each side.

9. Notes should be double-spaced. Triple-spacing should be used between notes.

10. Numbers should be lettered at a height of 3mm or 1/8" inch.

11. Fractions should be lettered at a height of 6mm. or 1/4" inch.

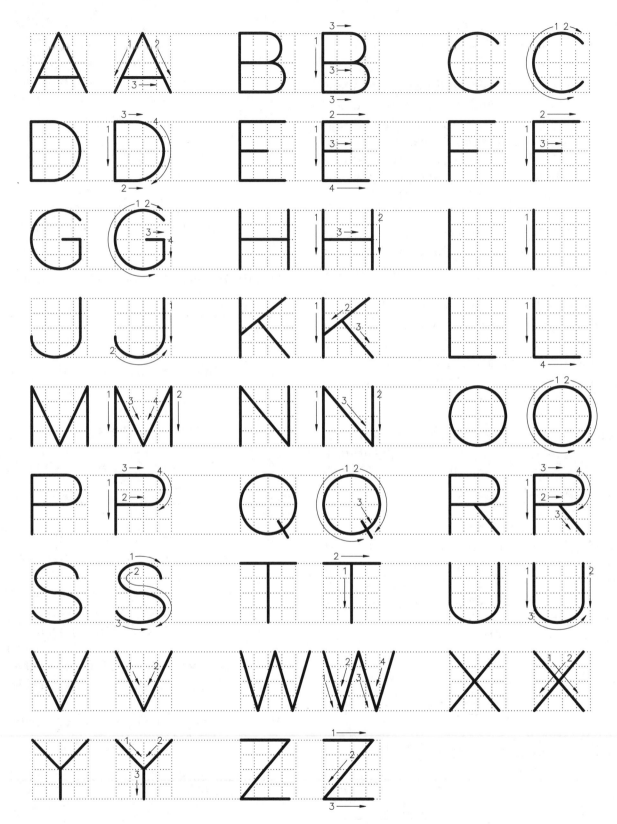

Figure LTR-2 — The typical individual strokes for each letter are illustrated. Notice that each letter (with the exception of "I") requires more than one stroke.

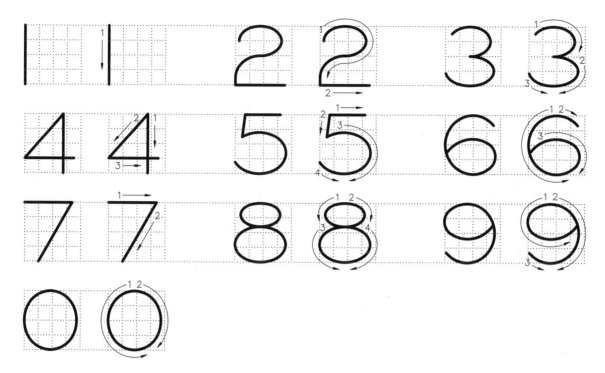

Figure LTR-3 — The typical individual strokes for each numeral are illustrated.

DRAW EACH LETTER TWICE USING THE GRID AS AN AID IN FORMING THE LETTERS.
CONCENTRATE ON MAKING EACH LETTER WITH SEPARATE AND DISTINCT STROKES.

A   B   C

D   E   F

G   H   I

J   K   L

M   N   O

P   Q   R

S   T   U

V   W   X

Y   Z

1   2   3

4   5   6

7   8   9

ENGINEERING GRAPHICS	NAME: _____ FILE NO.: _____ SECTION: _____	GRADE	LETTERING
			LTR~1

DRAW EACH LETTER TWICE USING THE SQUARE AS AN AID IN FORMING THE LETTERS.
CONCENTRATE ON MAKING EACH LETTER WITH SEPARATE AND DISTINCT STROKES.

A   B   C

D   E   F

G   H   I

J   K   L

M   N   O

P   Q   R

S   T   U

V   W   X

Y   Z

1   2   3

4   5   6

7   8   9

10   11   12

13   14   15

| ENGINEERING GRAPHICS | NAME:
 FILE NO.:   SECTION: | GRADE | LETTERING
 LTR~2 |

DRAW EACH LETTER FIVE TIMES USING THE GRID AS AN AID IN FORMING THE LETTERS.
CONCENTRATE ON MAKING EACH LETTER WITH SEPARATE AND DISTINCT STROKES.

B B B B B   O O O O O

R R R R R   G G G G G

J J J J J   P P P P P

A A A A A   K K K K K

N N N N N   U U U U U

M M M M M   D D D D D

F F F F F   W W W W W

S S S S S   E E E E E

REPEAT EACH LINE OF TEXT USING THE GUIDELINES.   THE VERTICAL LINES ARE INCLUDED AS
A GUIDE FOR DRAWING CERTAIN LETTERS.

REMOVE ALL SURFACE MARKS

REMOVE ALL SURFACE MARKS

INSTALL LOWER ASSEMBLY GUIDE

INSTALL LOWER ASSEMBLY GUIDE

PROVIDE ADDITIONAL POWER SOURCE

PROVIDE ADDITIONAL POWER SOURCE

SECURE BOOM AND CABLE

SECURE BOOM AND CABLE

| ENGINEERING GRAPHICS | NAME: | | GRADE | LETTERING |
| | FILE NO.: | SECTION: | | LTR~3 |

DRAW EACH LETTER FIVE TIMES USING THE GRID AS AN AID IN FORMING THE LETTERS.
CONCENTRATE ON MAKING EACH LETTER WITH SEPARATE AND DISTINCT STROKES.

C C C C C   H H H H H

L L L L L   Q Q Q Q Q

T T T T T   V V V V V

X X X X X   Y Y Y Y Y

2 2 2 2 2   3 3 3 3 3

4 4 4 4 4   5 5 5 5 5

6 6 6 6 6   7 7 7 7 7

8 8 8 8 8   9 9 9 9 9

REPEAT EACH LINE OF TEXT USING THE GUIDELINES.   THE VERTICAL LINES ARE INCLUDED AS
A GUIDE FOR DRAWING CERTAIN LETTERS.

CLOSE EVERY DRAIN HOLE
CLOSE EVERY DRAIN HOLE

INSERT EACH OF THE 68 SCREWS
INSERT EACH OF THE 68 SCREWS

LOCATE 29 SLOTS IN THE COVER
LOCATE 29 SLOTS IN THE COVER

45 EXTRA SCREWS ARE INCLUDED
45 EXTRA SCREWS ARE INCLUDED

ENGINEERING GRAPHICS	NAME:		GRADE	LETTERING
	FILE NO.:	SECTION:		LTR~4

Provide the missing measurements using a scale of 1 grid equal to 6 mm. Do not include the notation "mm" with each measurement. The "R" indicates a RADIUS and the "Ø" represents a DIAMETER measurement.

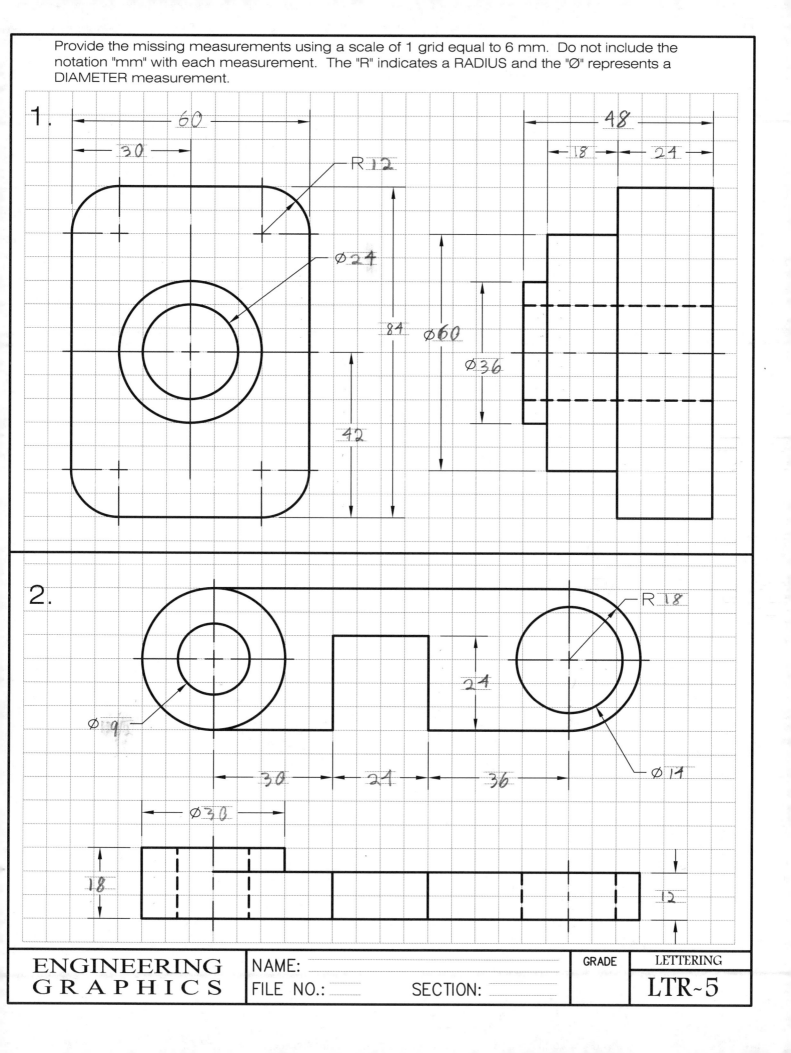

1.

60
30
R 12
Ø 24
48
18
24
84
Ø 60
Ø 36
42

2.

R 18
24
Ø 9
Ø 14
30
24
36
Ø 30
18
12

ENGINEERING GRAPHICS

NAME:
FILE NO.:   SECTION:

GRADE

LETTERING

LTR-5

SKETCHING

Sketching is a method of visualizing and conceptualizing your idea that allows you to communicate that idea with others. Sketches are not intended to be final engineering documents or drawings, but are a step in the process of going from "idea" to "final design". Sketches can be quickly executed to convey your conceptual ideas.

Use a wooden pencil with soft HB lead or a. mechanical pencil in 5mm, .7 mm or .9mm. It is easier to sketch with larger lead sizes (.7mm and .9mm). Erasures are simplified with a good quality white vinyl eraser.

SKETCHING LINES:

While sketching, it is important to visually mark the beginning point and ending point of the line with the understanding that your hand will always follow your eye. The grid system in this workbook is based on a scale of one grid equal to 3 mm or 1/8" and appears on each problem to facilitate sketching horizontal and vertical lines. For lines at an angle, reference the corners of the grid. When drawing vertical lines it is often easier to tilt the paper to a more comfortable angle. The four main types of lines and how they appear are illustrated in **(Figure SKT-1)**.

Construction Line

Object Line (Visible Line)

Hidden Line

Center Line

Section Line

Figure SKT-1 – Line types.

a. Construction Lines – are very light lines used to lay out the basic shape of the object. These lines are drawn only dark enough to be visible. Since construction lines will typically be left on the sketch and not erased, their appearance is more like a "ghost" line.

b. Object Lines – are dark, thick lines. Object lines show edges and figure outlines.

c. Hidden Lines – are thin and dashed. These lines represent edges of objects that are hidden or behind the face of the object.

d. Center Lines – are light lines with a short segment in the middle. Center lines extend through the center of the figure and beyond the edges of the figure.

e. Section Lines – are thin and angled. These lines represent the surface where the cutting plane slices through the object. These lines are lighter than an object line.

SKETCHING CIRCLES:

Sketching a circle requires the construction of reference points so that the circle appears round and uniform.

The steps illustrated in **Figure SKT-2** are:

a. Draw a square with construction lines using the diameter of the circle as the width and height of the square.

b. Divide the square into four equal squares using the center lines of the circle.

c. Draw a light construction line diagonally from each corner of the major square to the opposite corner.

d. Add four smaller diagonals to make a diamond crossing the two original diagonal lines. Mark a point half way between these four smaller diagonal lines (the diamond) and the outside corners of the major square. These marks become the target for drawing a circle and ensure that it appears round.

e. The circle is actually drawn with four smaller arcs. This allows you to turn the paper as you construct the circle.

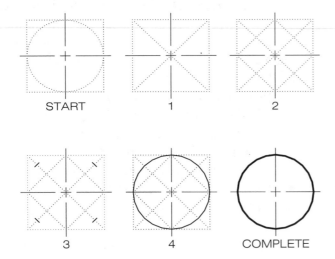

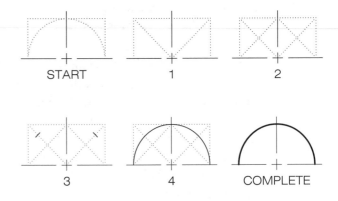

Figure SKT-2 – The geometric construction required to sketch a circle.

Figure SKT-3 – The geometric construction required to sketch an arc.

SKETCHING ARCS:

Arcs represent only part of a circle and therefore require only part of the construction of a circle. Use as many quadrants as necessary (as in the construction of a circle) to construct as large an arc as required.

The construction for a large arc (part of a circle) is illustrated in **Figure SKT-3**.

a. Layout the arc by noting its center point and its radius vertically and horizontally. Draw the necessary portion of the square to inscribe the arc.

b. Similar to the circle, draw diagonal lines from the center point to the opposite corners of the rectangle. Add the secondary diagonal lines to form "X's" within each square of the rectangle.

c. Add marks halfway between the second diagonal lines and the outer corners of the rectangle. These marks define the curve of the arc.

d. Starting on the edge of the vertical or horizontal point of tangency, begin drawing the arc while aiming to cross the mark made in the previous step and continuing to the next point of tangency.

ISOMETRIC SKETCHING:

Construction of an isometric drawing is covered in detail in the section of this workbook on ISOMETRICS.

To practice sketching an isometric, use the following rules:

a. Isometric sketches utilize planes parallel to the isometric axis.
b. Use the angled grid lines as a guide for the object lines that appear on the receding axes.
c. Transfer the measurements with one grid of the orthographic view equal to one grid vertically or one grid on either of the receding axes of the isometric view.

Isometric Ellipses:

As the object is rotated in order to see it as an isometric, the holes and all cylindrical features are also rotated. This rotation makes these circular features appear as ellipses instead of true circles as in **Figure SKT-5**.

The steps in drawing a simple isometric ellipse are:

a. Draw a parallelogram to scale that frames the completed ellipse. Notice that the orientation of the parallelogram will change based on the face of the isometric.

b. Find the mid-point of all four (4) sides of the parallelogram by measurement or by drawing a hidden line top to bottom and side to side and mark these intersections.

c. Draw an arc from the mid-point of the left side to the mid-point at the top of the rectangle. Repeat the arcs from the mid-point on one side to the mid-point of the adjacent side.

d. If necessary, trace over the ellipse to make it appear as one continuous line.

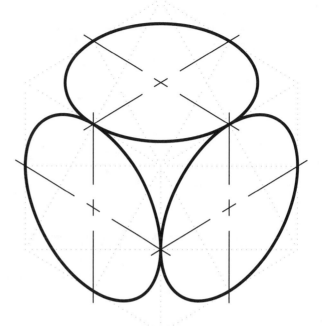

Figure SKT-5 – The ellipses are oriented to the parallelogram constructed in the face where they appear.

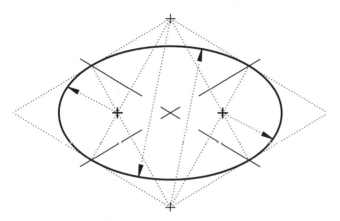

Figure SKT-4 – The ellipse is constructed from a parallelogram and drawn with four arcs.

The ellipses are oriented to the three main faces of the isometric box. This requires the rotation of the parallelogram as you move from one face to another. The orientation of the parallelogram relative to the three faces of the isometric box is shown in **Figure SKT-5**.

1. DRAW EACH LINE FOUR TIMES USING THE GUIDELINES. CONCENTRATE ON DRAWING EACH LINE TO MATCH THE GIVEN LINE IN LINE WEIGHT AND SPACING.

Object Line (Visible Line)

Hidden Line

Center Line

Section Line

2. REPEAT EACH LINE TYPE USING THE GUIDELINES AS AN AID IN SPACING AND LOCATION.

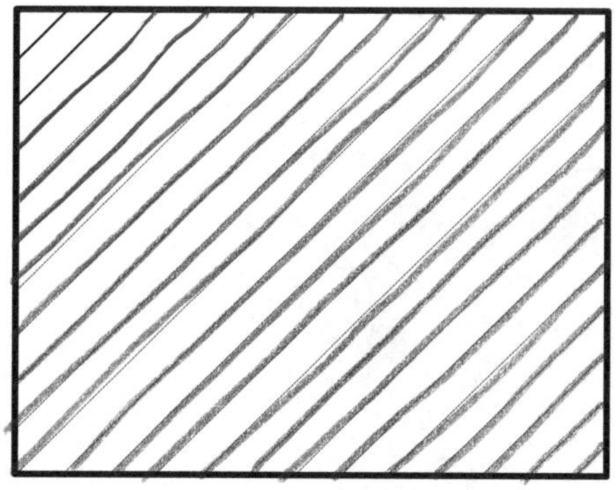

Section Lines (representing where a cutting plane has sliced through an object) are typically drawn at a 45° angle.

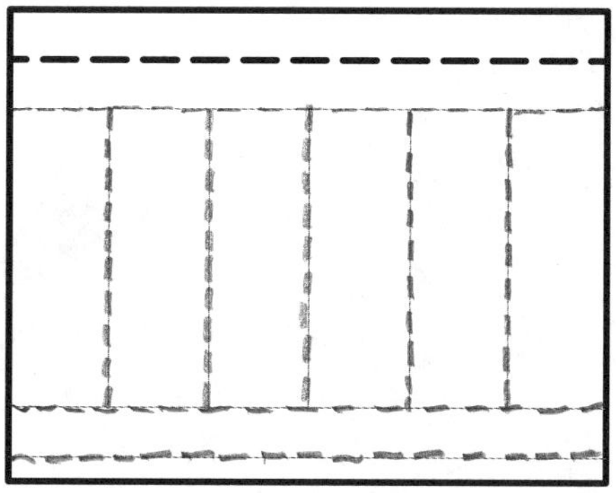

Hidden Lines (representing edges that are not visible)

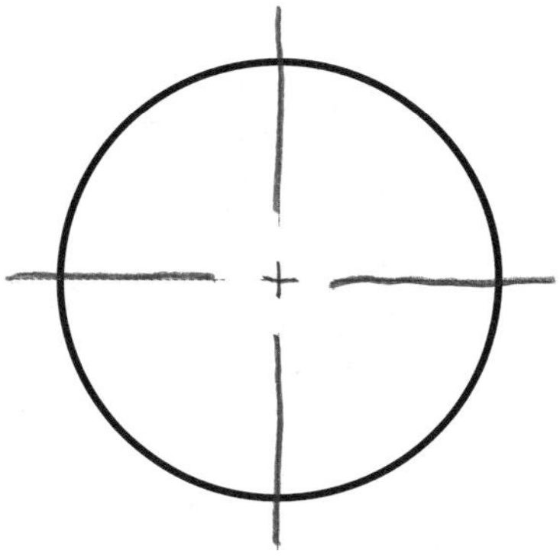

Center Lines (representing the center of a hole or cylinder) are positioned so that the center marks are in the center of the circle.

ENGINEERING GRAPHICS

NAME:

FILE NO.:   SECTION:

GRADE

SKETCHING

SKT~1

1. USING THE ILLUSTRATION BELOW OF THE CONSTRUCTION OF A CIRCLE CIRCUMSCRIBED WITHIN A SQUARE, DRAW CIRCLES IN THE TWO LARGE SQUARES BELOW. LIGHTLY SKETCH THE CONSTRUCTION LINES TO LAYOUT THE CIRCLE IN THE SECOND SQUARE.

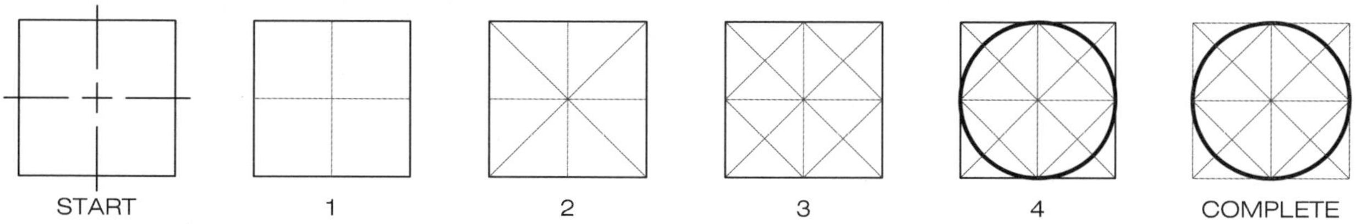

START   1   2   3   4   COMPLETE

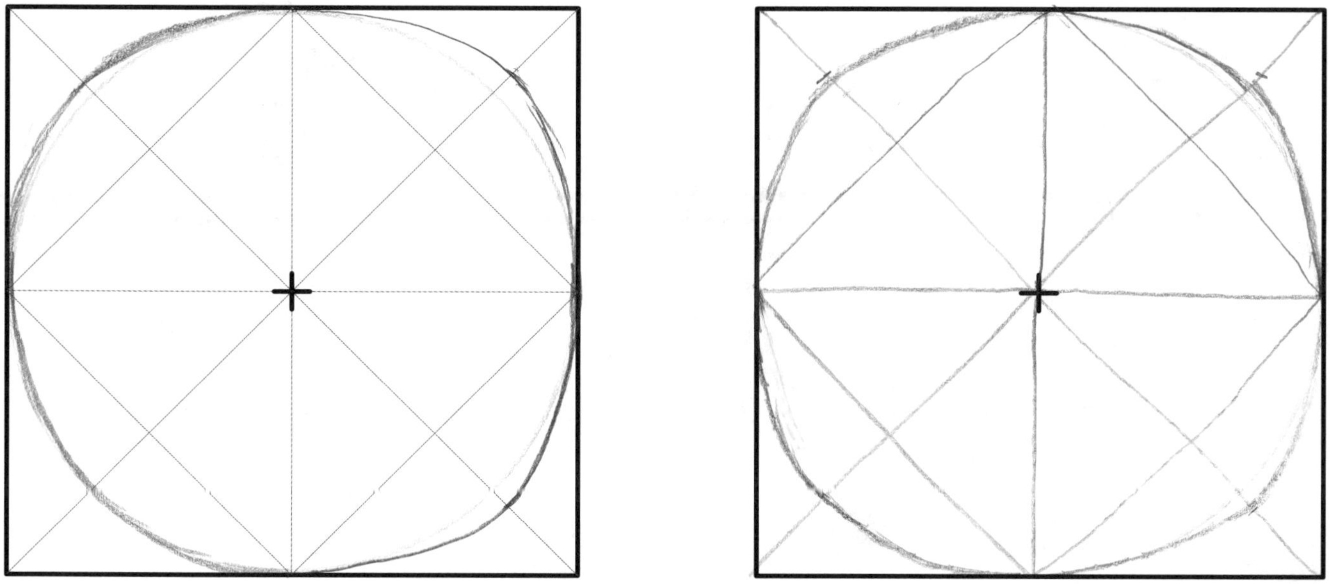

2. DRAW TWO ARCS IN THE SQUARES BELOW. DRAW THE CONSTRUCTION LINES TO LAYOUT THE ARC IN THE SECOND SQUARE. NOTE THAT YOU ARE DRAWING ONLY A QUARTER OF A CIRCLE.

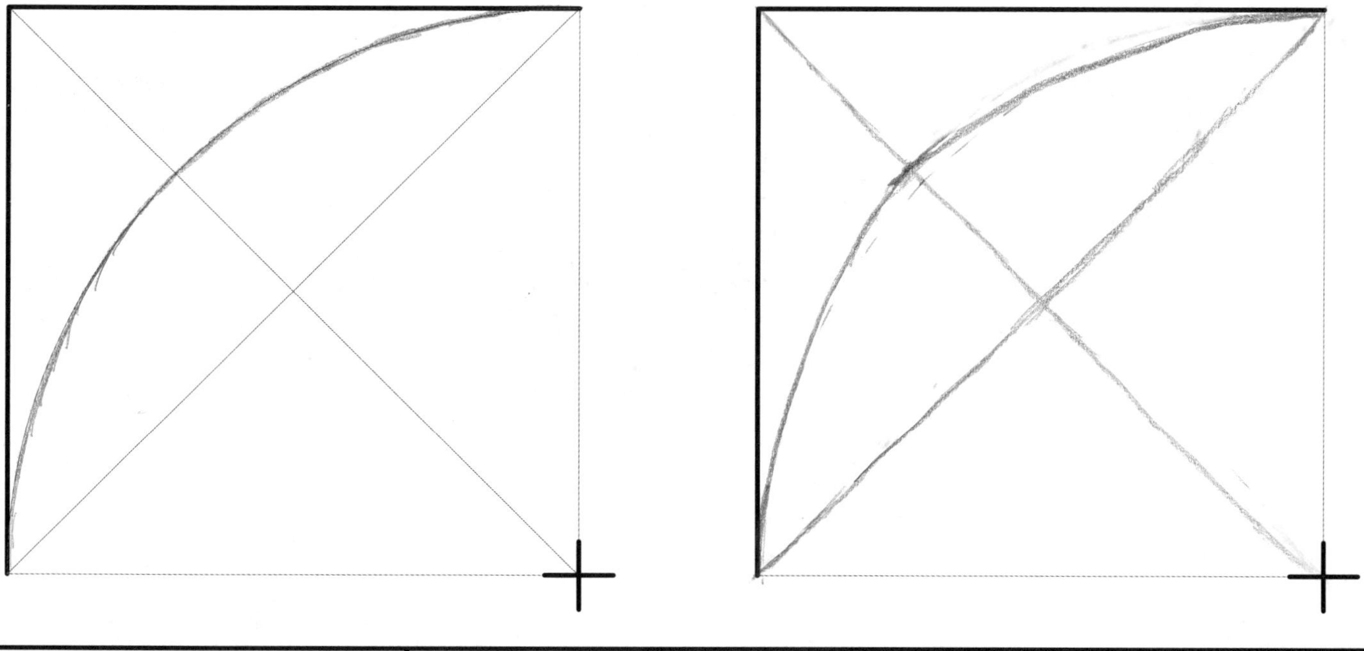

ENGINEERING
GRAPHICS

NAME:
FILE NO.:   SECTION:

GRADE

SKETCHING

SKT~2

1. DRAW A FREEHAND SKETCH OF THE GIVEN
OBJECT WITH REFERENCE TO CORNER "A" AS
THE STARTING CORNER. TRANSFER
MEASUREMENTS WITH ONE GRID OF THE
GIVEN DRAWING EQUAL TO ONE GRID OF THE
NEW SKETCH. NOTICE THAT YOUR SKETCH
WILL BE TWICE AS LARGE AS THE GIVEN
SKETCH.

2. DRAW A FREEHAND SKETCH OF THE GIVEN
OBJECT WITH REFERENCE TO CORNER "A" AS
THE STARTING CORNER. TRANSFER
MEASUREMENTS WITH ONE GRID OF THE
GIVEN DRAWING EQUAL TO ONE GRID OF THE
NEW SKETCH. LAYOUT THE CONSTRUCTION
LINES FOR THE ROUNDED ENDS FIRST.

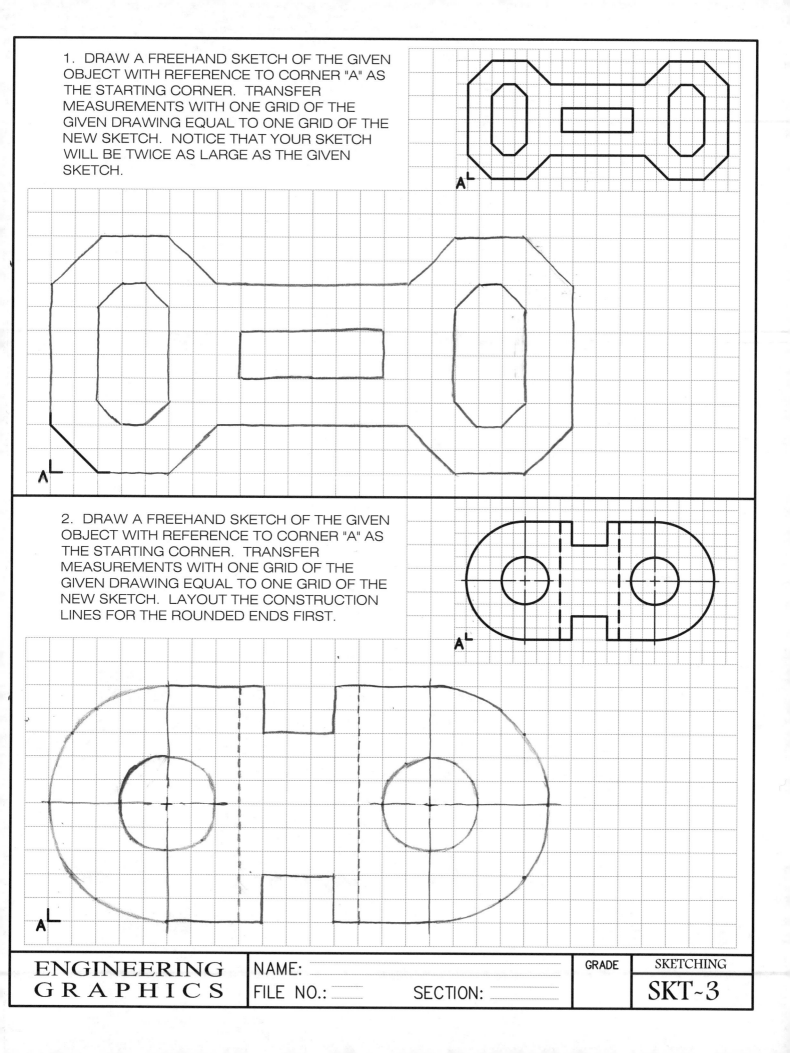

ENGINEERING
GRAPHICS

NAME:

FILE NO.:   SECTION:

GRADE

SKETCHING

SKT~3

1. DRAW A FREEHAND SKETCH OF THE GIVEN OBJECT.

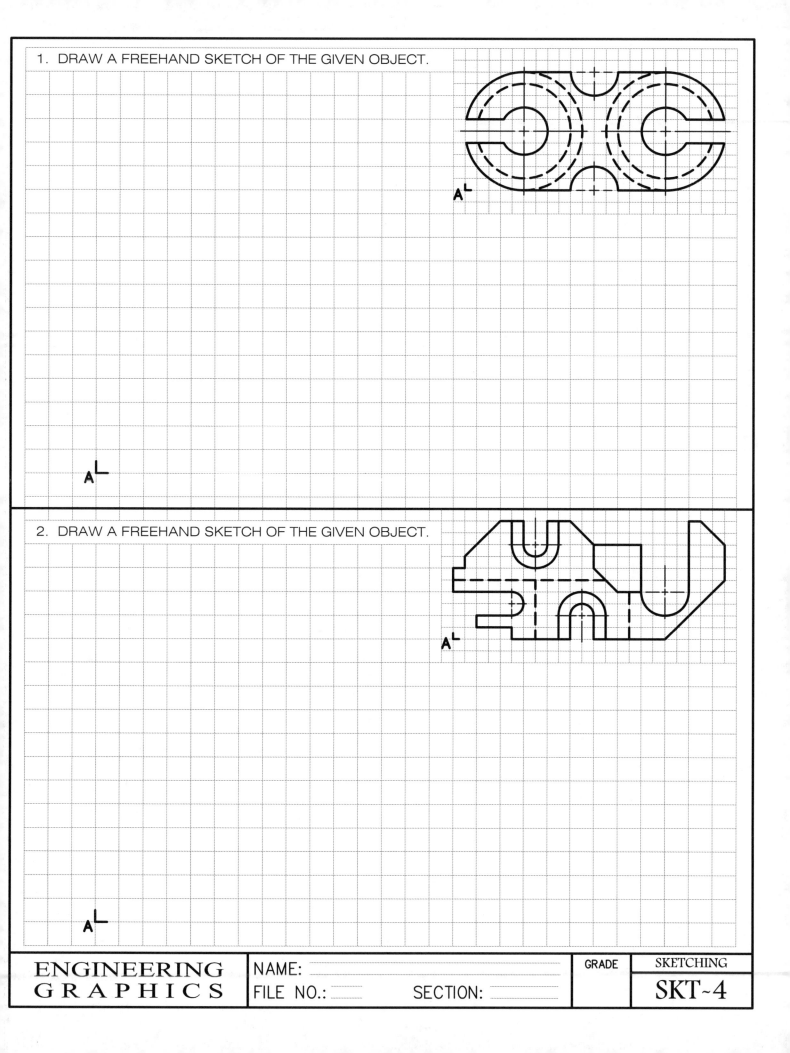

A⌐

A⌐

2. DRAW A FREEHAND SKETCH OF THE GIVEN OBJECT.

A⌐

A⌐

ENGINEERING
GRAPHICS

NAME:
FILE NO.:   SECTION:

GRADE

SKETCHING

SKT~4

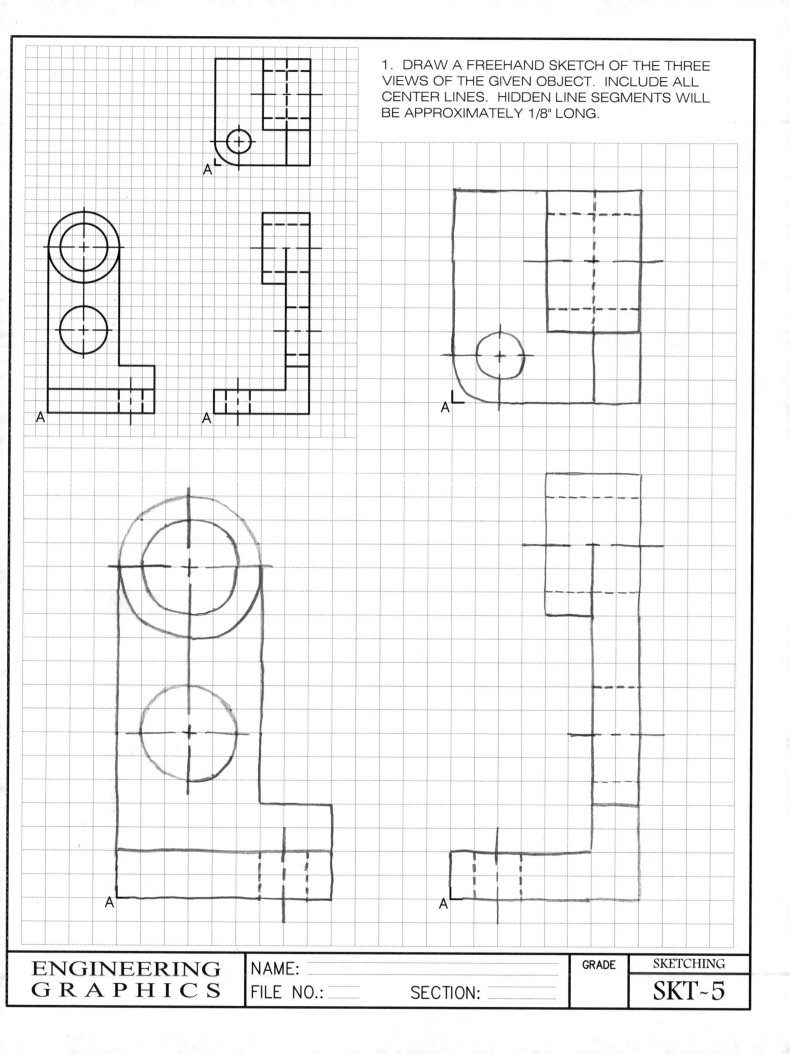

1. DRAW A FREEHAND SKETCH OF THE THREE VIEWS OF THE GIVEN OBJECT. INCLUDE ALL CENTER LINES. HIDDEN LINE SEGMENTS WILL BE APPROXIMATELY 1/8" LONG.

A

A

A

A

A

ENGINEERING GRAPHICS

NAME:

FILE NO.:

SECTION:

GRADE

SKETCHING

SKT~5

1. Using the EXAMPLE below, construct the four-center ELIPSES for Problems A and B. Refer to the EXAMPLE to establish the radius points for Problem B.

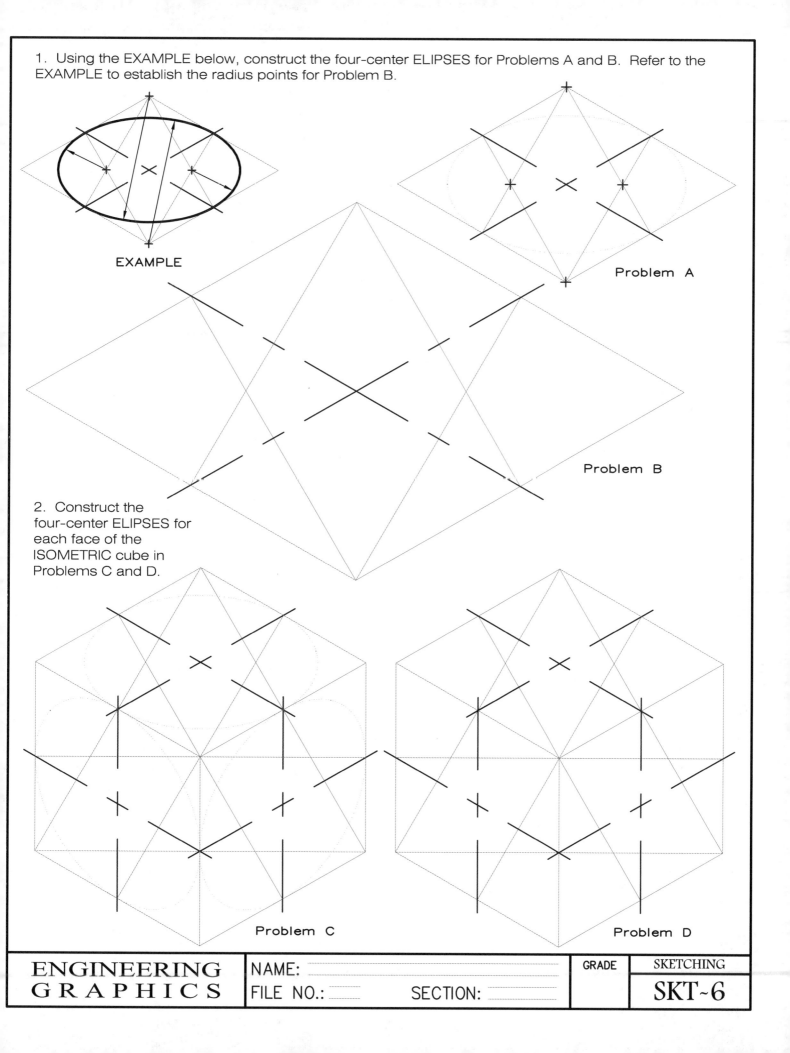

EXAMPLE

Problem A

Problem B

2. Construct the four-center ELIPSES for each face of the ISOMETRIC cube in Problems C and D.

Problem C

Problem D

ENGINEERING
GRAPHICS

NAME: _____

FILE NO.: _____ SECTION: _____

GRADE

SKETCHING

SKT~6

Using the technique for drawing a four-center ELLIPSE, and referring to the EXAMPLE, sketch the cylinders indicated in each problem. Note that the visible end is marked and the diagonal lines that determine the location of the visible edges are given.

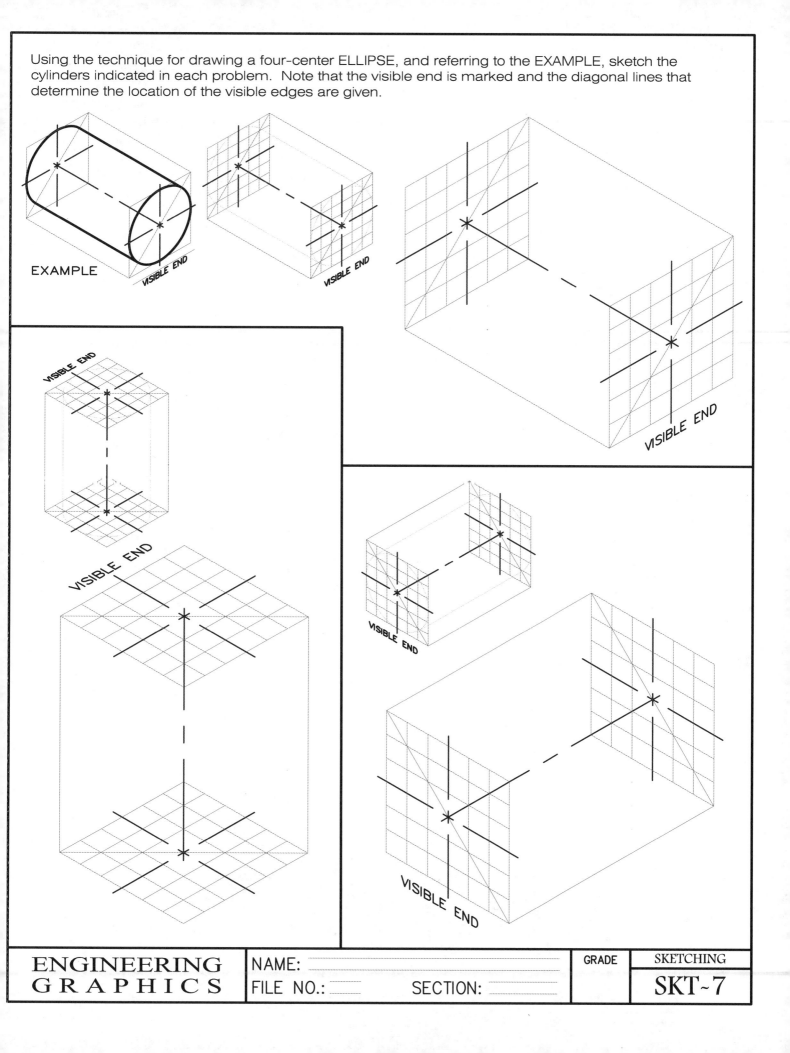

EXAMPLE

VISIBLE END

VISIBLE END

VISIBLE END

VISIBLE END

VISIBLE END

VISIBLE END

VISIBLE END

ENGINEERING
GRAPHICS

NAME:
FILE NO.:   SECTION:

GRADE

SKETCHING

SKT~7

Sketch the object drawn in the ISOMETRIC view on the ISOMETRIC grid provided.

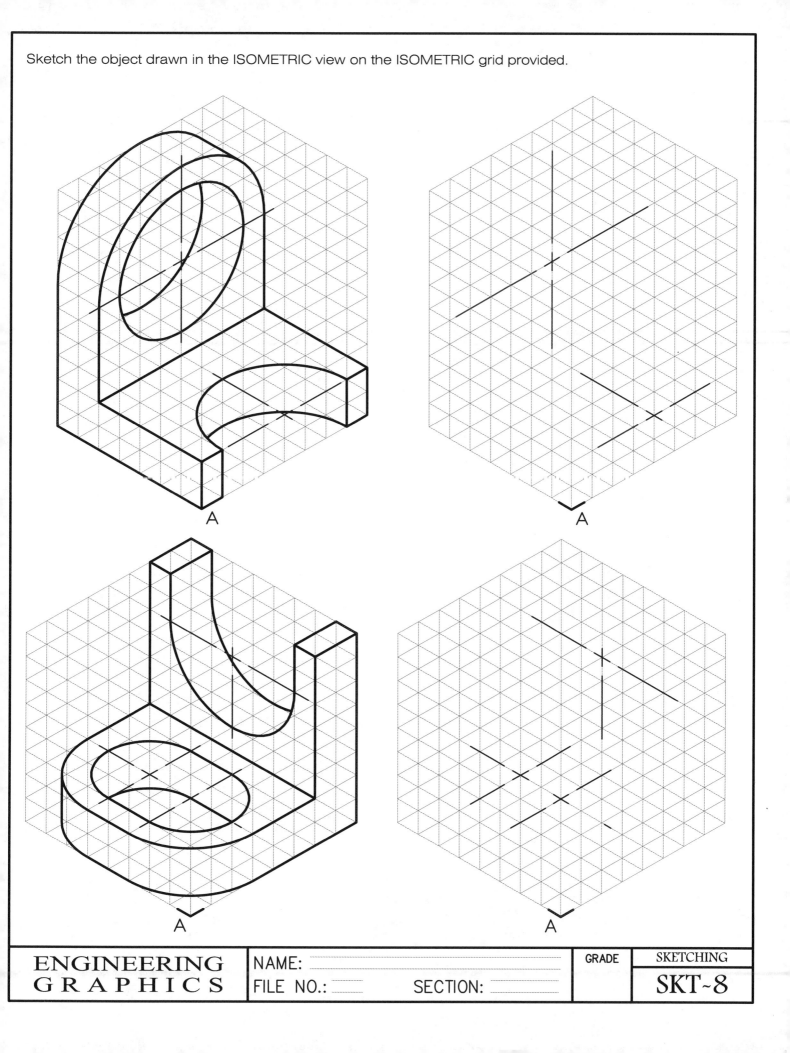

A

A

A

A

ENGINEERING
GRAPHICS

NAME:

FILE NO.:   SECTION:

GRADE

SKETCHING

SKT~8

ORTHOGRAPHIC PROJECTION

Orthographic projection is the organization of multiple views of an object. In order to fully and accurately represent the object, more than one view is typically required. As shown in **Figure ORT-1**, the automobile requires at least three views to fully represent the object.

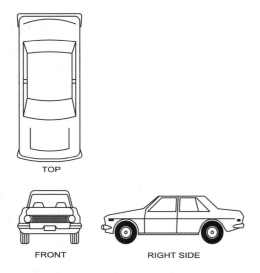

TOP

FRONT   RIGHT SIDE

Figure ORT-1 – Three views are necessary to describe the size and shape of the automobile.

Orthographic projections are two dimensional (2D) views of three dimensional (3D) objects. Orthographic projections depict an object that is viewed along parallel lines that are perpendicular to the object. These lines remain parallel to the projection plane and are not convergent. All orthographic projections are composed of six sides or planes of projection. All six are not necessary to describe the object.

These projections represent the object's height, width and depth. Notice that each one of the six views represents only two dimensions of the object. The six views of an orthographic projection are:

Top, Bottom, Left, Right, Front and Back

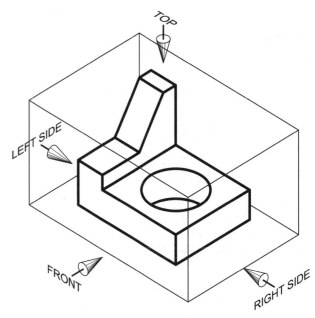

Figure ORT-2 – The views are named by the viewing direction of the object.

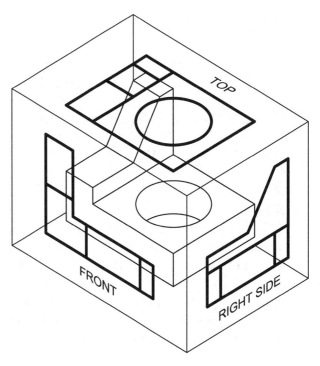

Figure ORT-3 – The lines of the object are projected perpendicular to the face to create each view.

a. Top and Bottom views show DEPTH and WIDTH.

b. Left and Right views show HEIGHT and DEPTH

c. Front and rear views show HEIGHT and WIDTH.

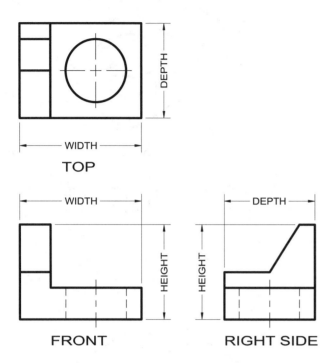

Figure ORT-4 – Each view provides only two measurements. However, by using any two views, the three measurements of height, width and depth are provided.

No orthographic view can show height, width and depth in the same view. Each view only depicts two dimensions. Therefore, a minimum of two projections (or views) are required to display all three dimensions of an object. Typically, most orthographic drawings use three views to accurately depict the object unless more are required for clarity.

When drawing orthographic projections, spacing is usually equal between each of the projections. However, this equal distance is not mandatory. The Front, Top, and Right Side views are most frequently used to depict an orthographic projection. However, choose the views with the least number of hidden lines for better visualization of the object.

In the system of projection (Third Angle) used in the United States, the top view is always placed over the front view and therefore the front view is always under the top view.

When transferring measurements between views, the width measurement can be projected from the front view up to the top view or vice versa and that the height measurement can be easily projected directly across from the front view to the right side or the left side view of the object or vice versa. Depth measurements are transferred from the top view to the right or left side views or vice versa. The depth measurements can be easily transferred by counting grids.

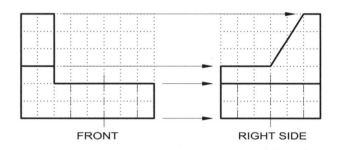

Figure ORT-5 – Height measurements are easily projected between the two views using the grids.

Another method to transfer depth between the top and right side (or left side) views, is drawing a mitre or 45° angle. To transfer a depth measurement from the right side view to the top view, a vertical line is drawn from the front corner of the right view until it intersects the horizontal line drawn from the front corner of the top view. The most forward edges of the top and right side views are the edges closest to the front view. The mitre is located where these two construction lines cross. Each line is transferred vertically or horizontally to the mitre line and then directed at a 90° angle to the other view.

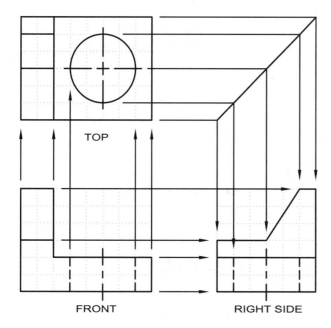

Figure ORT-6 – The mitre line (drawn at a 45° angle) is used to transfer depth measurements between the top and right side (or left side) views.

Using the mitre line allows you to check and verify that the depth indicated in each view is the same.

PLANES AND SURFACES

Each type of plane has unique characteristics when viewed in orthographic projection. In order to understand the three basic types of planes, each is presented with an illustration to indicate graphically how they vary. It is important to know the differences between the *normal, inclined* and *oblique planes* to understand how they will appear in orthographic projection.

a. Normal Planes will appear as an edge in two views and a true sized plan in the remaining view when using three views such as a top, front and right side. A *normal plane* would appear as an edge in four views and a true sized plane in two views if it was projected to all six possible views in orthographic projection. Normal planes are illustrated in **Figure ORT-7**.

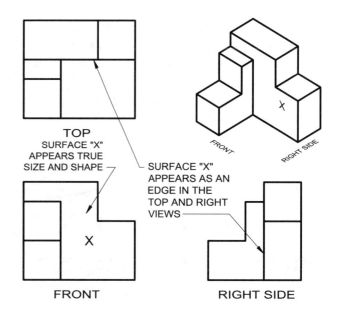

Figure ORT-7 – Normal Plane

b. Inclined Planes will appear as an edge view in only one of the three views. The inclined plane will appear as a rectangular surface in the other two views. Although the two rectangular surfaces appear "normal" they are in fact "foreshortened" and do not appear true size and shape. An *inclined plane* appears in **Figure ORT-8**.

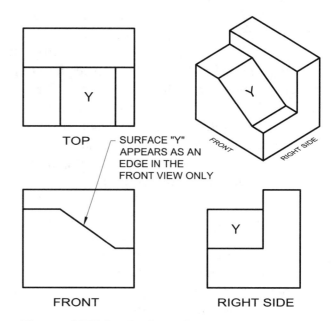

Figure ORT-8 – Inclined Plane

c. Oblique Planes will not appear as an edge view in any of the six views since they are not parallel or perpendicular to the projection planes. They always appear as a "plane" and have the same number of corners in each of the six views. **Figure ORT-9** illustrates an *oblique plane*.

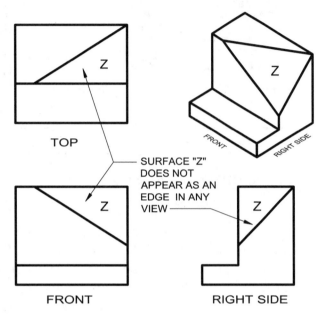

Figure ORT-9 – Oblique Plane

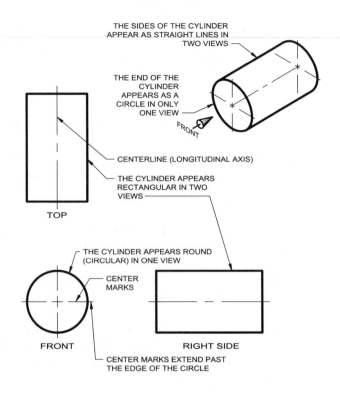

Figure ORT-10 – A cylinder only appears circular in one of the three orthographic views.

CURVED SURFACES

A cylinder will appear as a circle in one view and a rectangular shape the other two views. A centerline is used to identify the *longitudinal axis* from the top to the bottom of the cylinder. The axis appears where the cylinder appears rectangular. Center marks are used to identify the center of the cylinder where it appears circular. The relationship between the isometric and orthographic views is illustrated in **Figure ORT-10**.

Holes, which are negative cylinders, follow the same rules as a cylinder since they will appear circular in only one view and rectangular in two views. However, since the edges of the hole are inside the material, hidden lines are used to represent the edges (or sides) of the hole. **Figure ORT-11** shows a simple hole in a block of material.

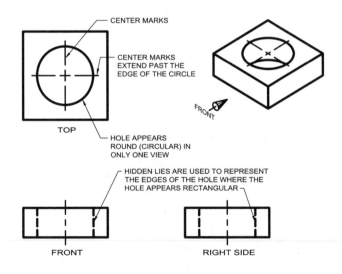

Figure ORT-11 – A hole is represented with hidden lines in the rectangular views.

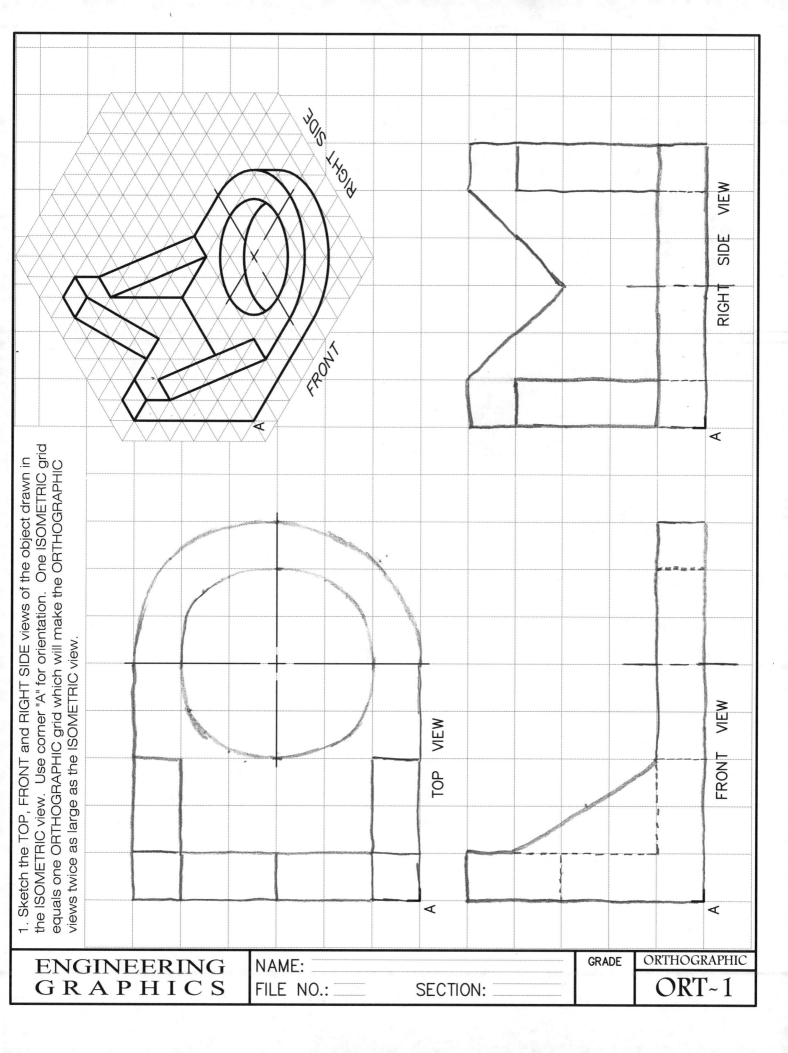

1. Sketch the TOP, FRONT and RIGHT SIDE views of the object drawn in the ISOMETRIC view. Use corner "A" for orientation. One ISOMETRIC grid equals one ORTHOGRAPHIC grid which will make the ORTHOGRAPHIC views twice as large as the ISOMETRIC view.

RIGHT SIDE

FRONT

A

RIGHT SIDE VIEW

A

TOP VIEW

A

FRONT VIEW

A

ENGINEERING
GRAPHICS

NAME: _____

FILE NO.: _____   SECTION: _____

GRADE

ORTHOGRAPHIC

ORT~1

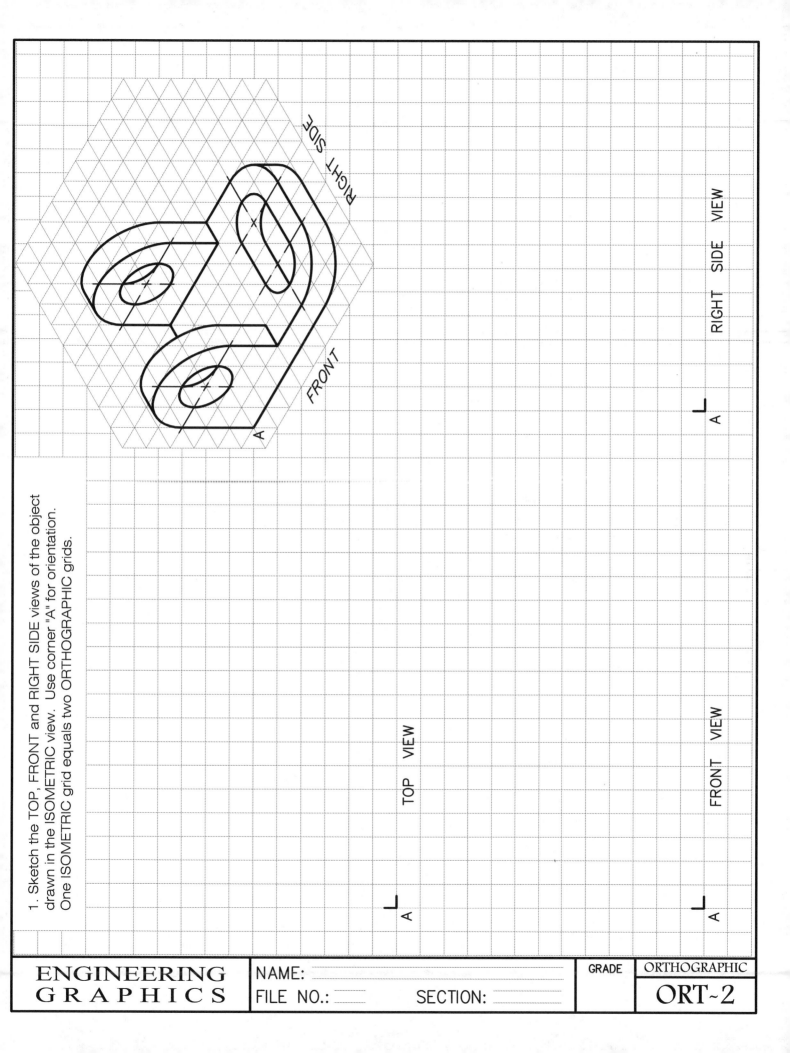

1. Sketch the TOP, FRONT and RIGHT SIDE views of the object drawn in the ISOMETRIC view. Use corner "A" for orientation. One ISOMETRIC grid equals two ORTHOGRAPHIC grids.

RIGHT SIDE

FRONT

A

TOP VIEW

A

RIGHT SIDE VIEW

A

FRONT VIEW

A

ENGINEERING GRAPHICS

NAME: _____

FILE NO.: _____ SECTION: _____

GRADE

ORTHOGRAPHIC

ORT~2

Sketch the ORTHOGRAPHIC views based on the ISOMETRIC drawings.

1.

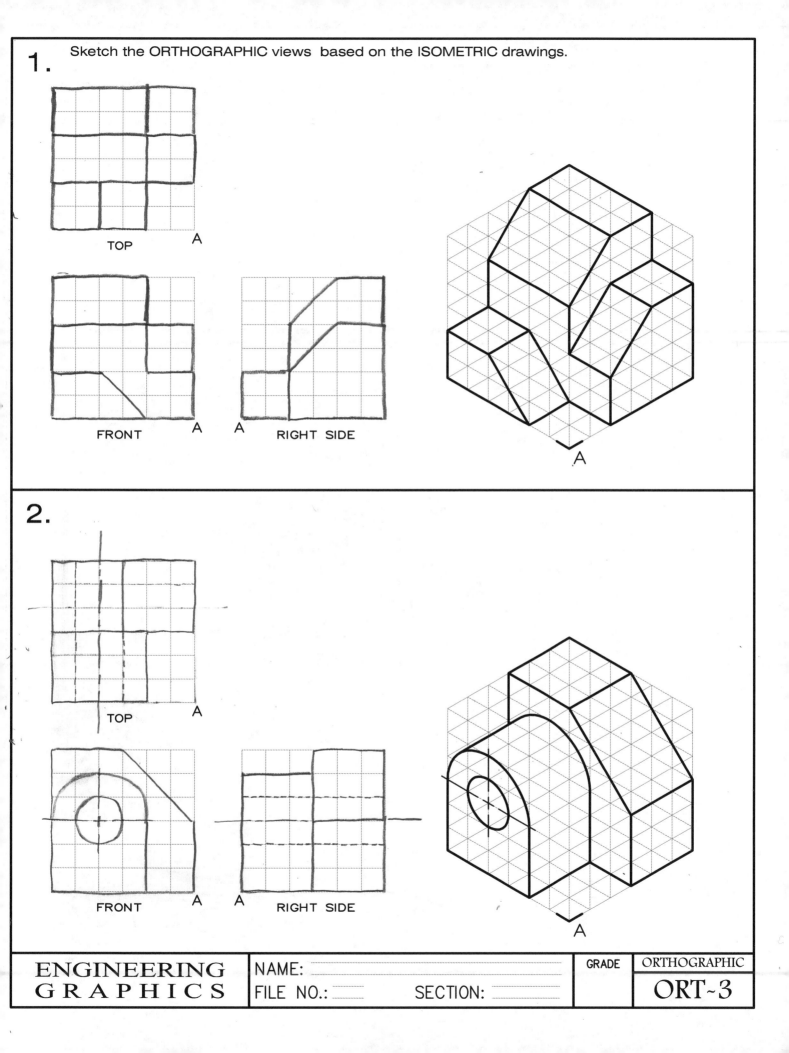

TOP

A

FRONT   A

A   RIGHT SIDE

A

2.

TOP   A

FRONT   A

A   RIGHT SIDE

A

ENGINEERING
GRAPHICS

NAME: _____
FILE NO.: _____   SECTION: _____

GRADE

ORTHOGRAPHIC

ORT~3

1.

Sketch the ORTHOGRAPHIC views based on the ISOMETRIC drawing.

TOP A

FRONT A   A   RIGHT SIDE

A

2.

TOP A

FRONT A   A   RIGHT SIDE

A

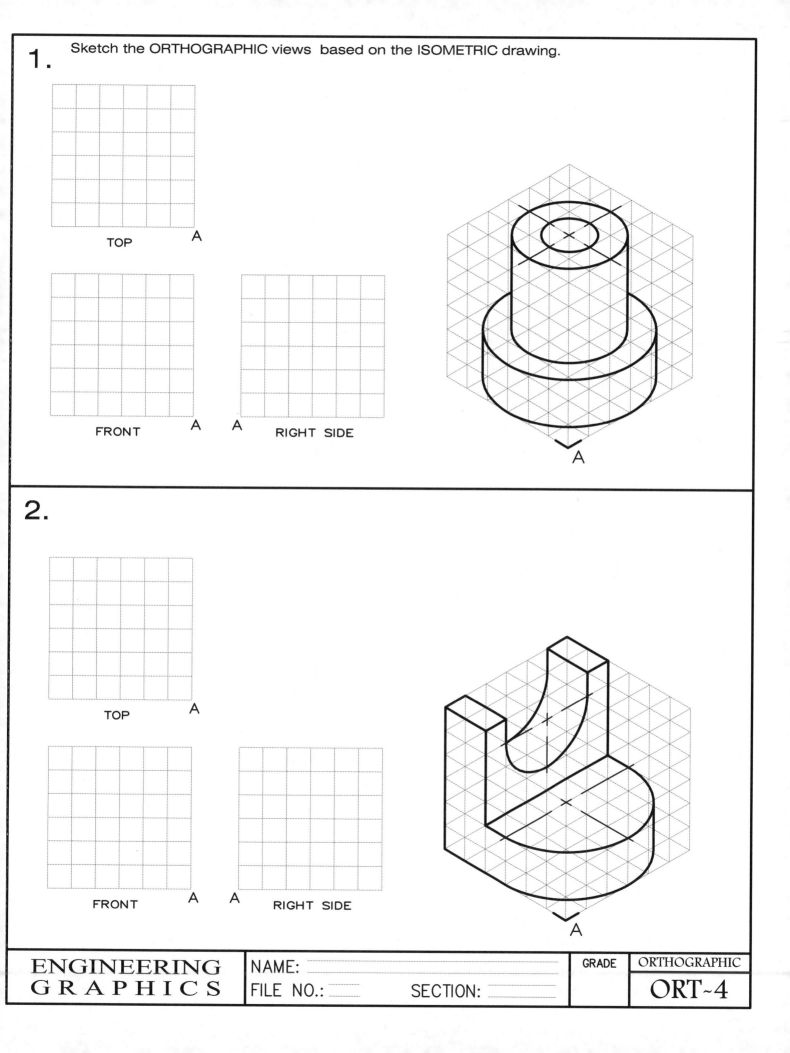

ENGINEERING
GRAPHICS

NAME: _____
FILE NO.: _____ SECTION: _____

GRADE | ORTHOGRAPHIC

ORT~4

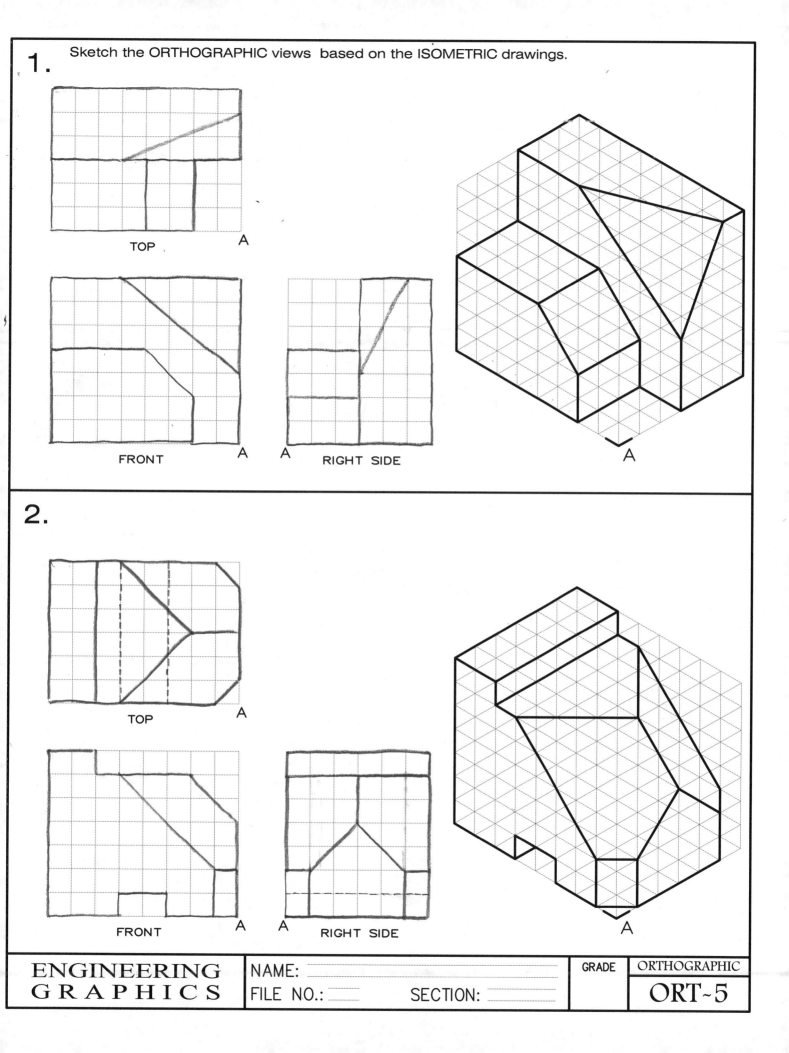

Sketch the ORTHOGRAPHIC views based on the ISOMETRIC drawings.

1.

TOP
A

FRONT
A

A
RIGHT SIDE

A

2.

TOP
A

FRONT
A

A
RIGHT SIDE

A

ENGINEERING
GRAPHICS

NAME: _____
FILE NO.: _____ SECTION: _____

GRADE | ORTHOGRAPHIC

ORT~5

Sketch the FRONT view for each problem. The given views are complete.

1.

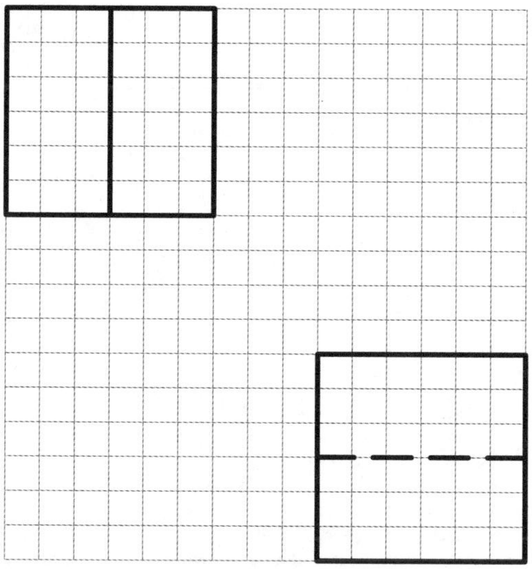

2.

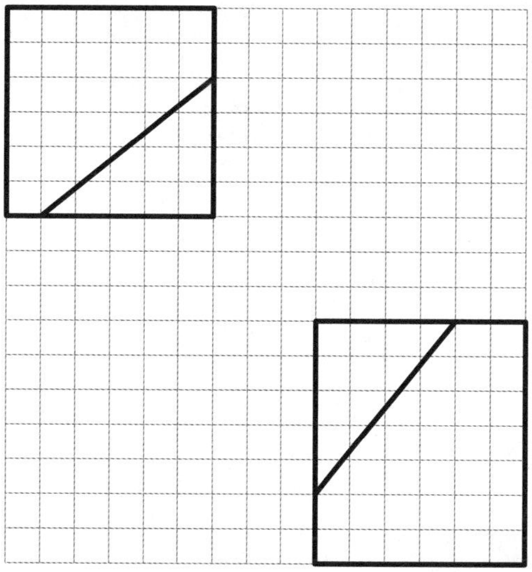

3.

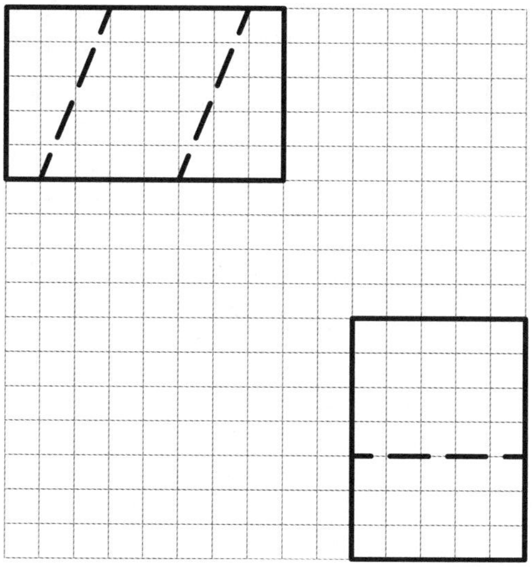

4.

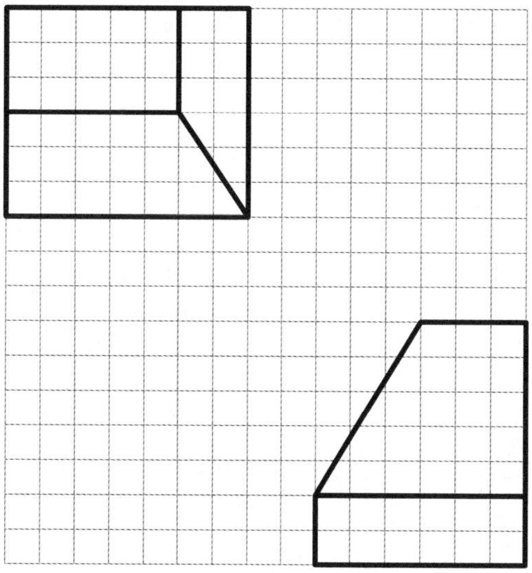

5.

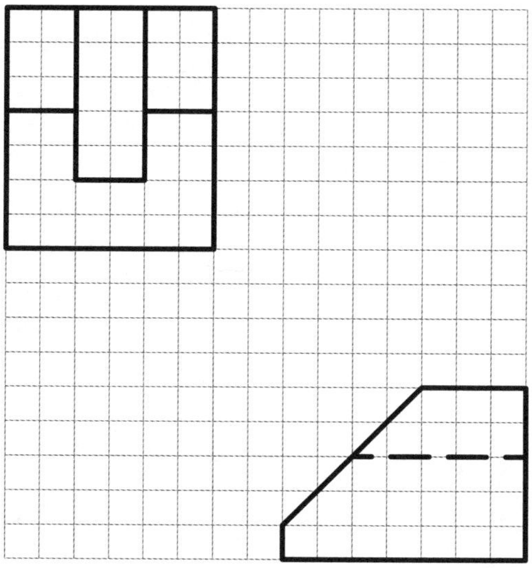

6.

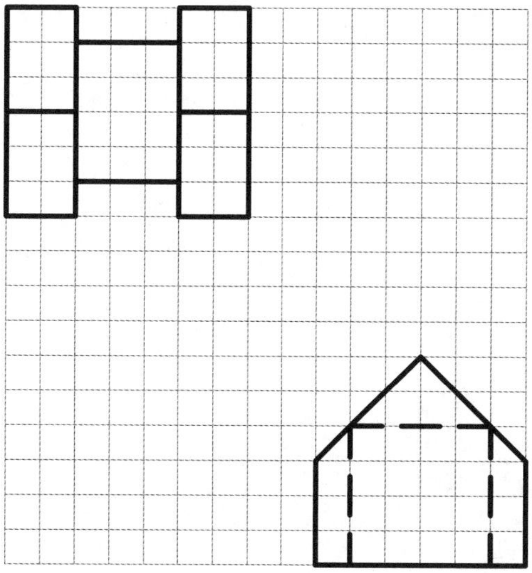

ENGINEERING
GRAPHICS

NAME: _____

FILE NO.: _____ SECTION: _____

GRADE

ORTHOGRAPHIC

ORT~6

The two given ORTHOGRAPHIC views are complete. Sketch the missing view.

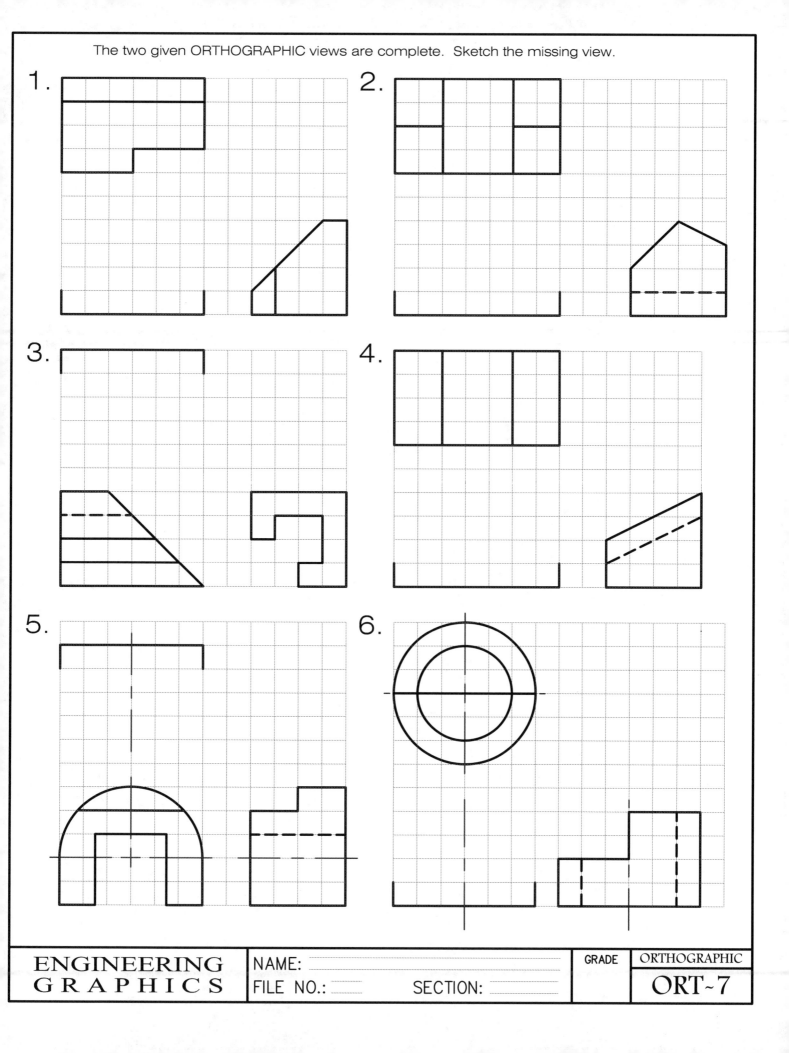

1.

2.

3.

4.

5.

6.

ENGINEERING
GRAPHICS

NAME: _____

FILE NO.: _____   SECTION: _____

GRADE | ORTHOGRAPHIC

ORT~7

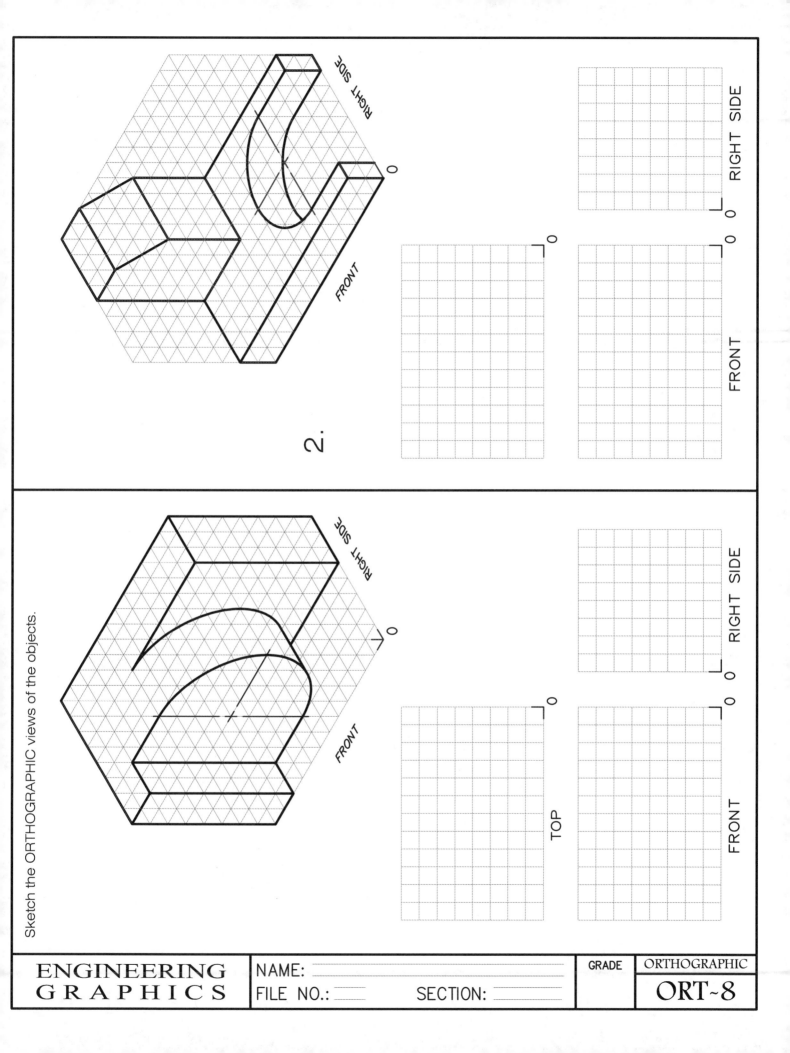

Sketch the ORTHOGRAPHIC views of the objects.

2.

RIGHT SIDE

FRONT

RIGHT SIDE

FRONT

TOP

ENGINEERING GRAPHICS

NAME:

FILE NO.:   SECTION:

GRADE

ORTHOGRAPHIC

ORT~8

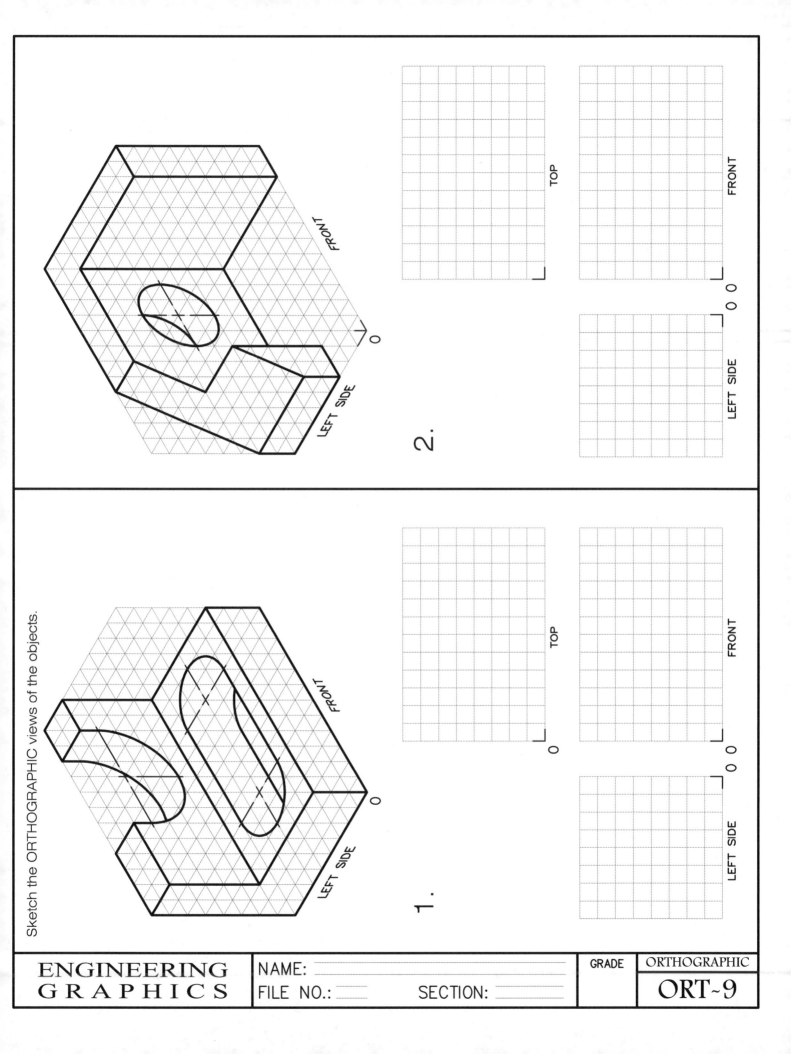

Sketch the ORTHOGRAPHIC views of the objects.

FRONT

LEFT SIDE

0

2.

TOP

FRONT

0 0

LEFT SIDE

FRONT

LEFT SIDE

0

1.

TOP

0

FRONT

0 0

LEFT SIDE

ENGINEERING
GRAPHICS

NAME: _____
FILE NO.: _____ SECTION: _____

GRADE

ORTHOGRAPHIC

ORT~9

Sketch the TOP, FRONT and LEFT PROFILE views of the object drawn in the ISOMETRIC view.

TOP

FRONT

LEFT PROFILE

A

A A

A

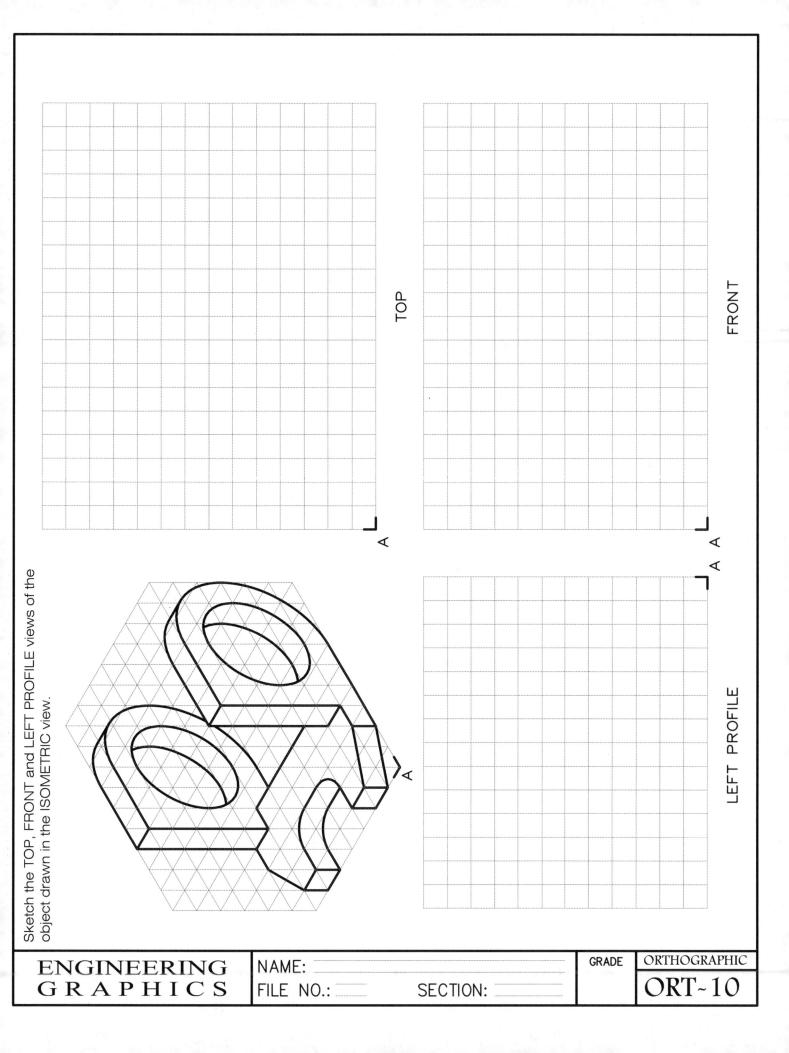

ENGINEERING
GRAPHICS

NAME:

FILE NO.:   SECTION:

GRADE

ORTHOGRAPHIC

ORT~10

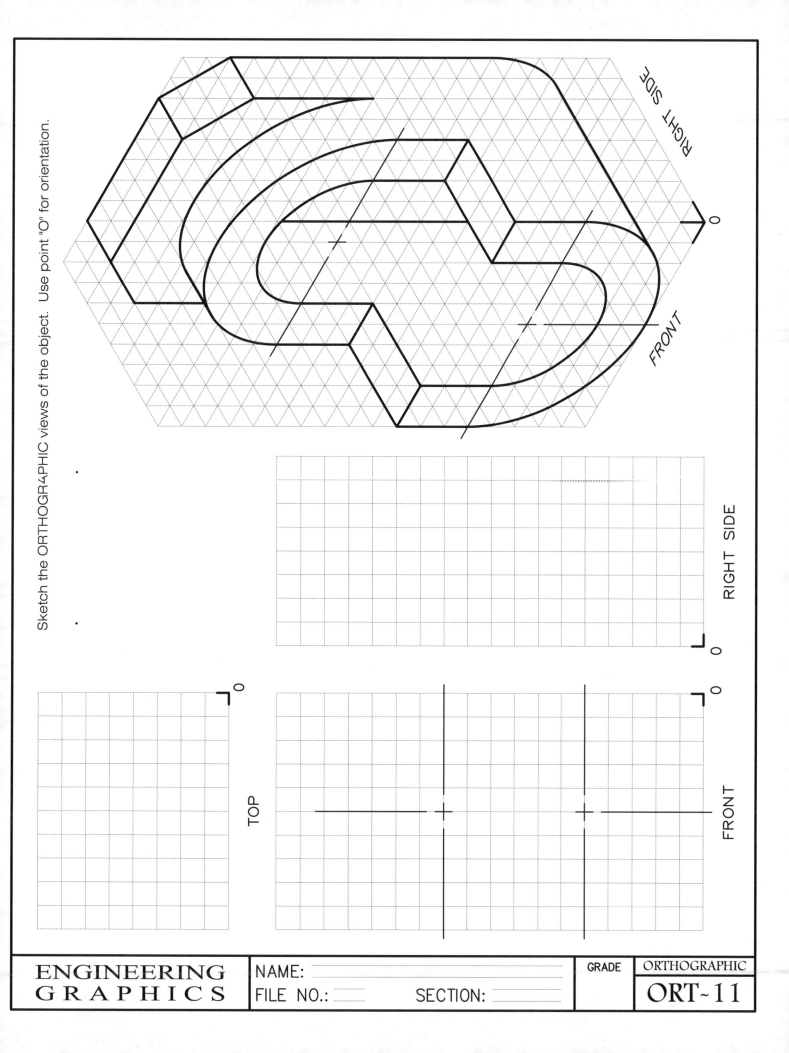

Sketch the ORTHOGRAPHIC views of the object. Use point "O" for orientation.

RIGHT SIDE

FRONT

O

RIGHT SIDE

O

TOP

O

FRONT

O

ENGINEERING
GRAPHICS

NAME:

FILE NO.:   SECTION:

GRADE

ORTHOGRAPHIC

ORT~11

ISOMETRIC DRAWINGS

Orthographic projection is used to describe an object with two dimensions in each view. In contrast, isometric drawings are valuable since they allow us to see all three dimensions at one time. This pictorial drawing provides a clearer representation of the object and makes it easier to visualize the object and understand the relationship of its various parts.

Isometric means "equal in measure" and refers to the fact that the three receding axes are tilted at 30°. Isometric drawings are constructed with parallel, non-converging lines which are drawn in exact proportion to render a three dimensional representation of an object. This is in contrast to orthographic projections which are drawn on a horizontal plane. Isometric drawings are constructed on a 30° incline to the projection plane.

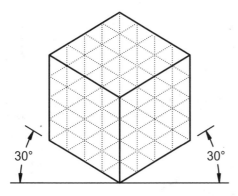

Figure ISO-1 – The isometric cube

The isometric problems in this workbook rely on an isometric grid for construction. The vertical and angled lines relate to the axes of the isometric cube. The measurements of the orthographic projection are transferred directly to the isometric grid (except in the case of inclined or oblique surfaces). For inclined or oblique surfaces, the corners are located and the lines are then connected from corner to corner. This technique is used because the angle that appears in the orthographic projection cannot be transferred directly to the isometric drawing.

ISOMETRIC DRAWING OF AN OBJECT WITH AN INCLINED SURFACE

The orthographic drawing shown in **Figure ISO-2** will be drawn as an isometric drawing in **Figure ISO-3**.

Notice the orientation of the top, front and right sides of the isometric grid. Because the object in the isometric is rotated and tilted, the isometric drawing is actually smaller than the orthographic views. To simplify the drawing of the isometric, measurements are transferred with one grid on the orthographic drawing equal to one grid on the isometric drawing.

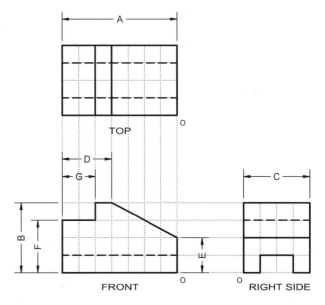

Figure ISO-2 – Orthographic drawing of an object with an inclined surface.

Step 1 - Begin by sketching the object as if it were a complete cube without any cuts. This defines the basic shape of the object. The measurements of overall Width (A), Height (B) and Depth (C) are transferred from the orthographic to the isometric drawing.

Step 2 - Next, make the inclined cut as shown in Step 2. Since the angle of the inclined surface cannot be transferred directly, you must locate the corners of the inclined surface and then draw lines to connect the corners. Notice that edges that are parallel in the orthographic views will also be parallel in the isometric drawing. This observation will simplify and quicken the construction of the isometric.

Step 3 - Add the rectangular cut across the left top edge. Notice that the rear edge of the cut disappears behind the raised portion of the block.

Step 4 – Add the slot that is cut into the bottom of the block on the right side. Only one receding axis of the slot will be visible.

ISOMETRIC DRAWING OF AN OBJECT WITH AN OBLIQUE SURFACE

The oblique surface shown in **Figure ISO-4** is drawn as an isometric in **Figure ISO-5**. The measurements J, K and L are transferred to the isometric grid to locate the corners of the oblique plane. The technique remains the same if the plane cuts through the original rectangular prism at an angle to produce a plane with four or more corners,. Each corner is transferred to the isometric and the lines between the corners are drawn.

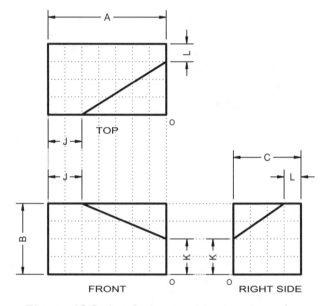

Figure ISO-4 – Orthographic drawing of an object with an oblique surface.

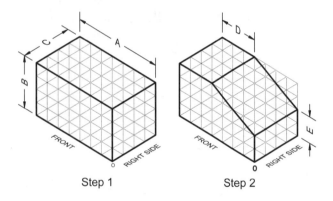

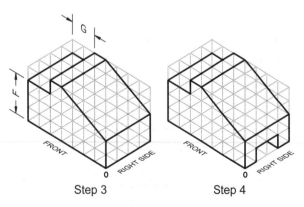

Figure ISO-3 –The isometric drawing is completed in a series of steps.

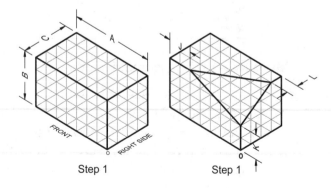

Figure ISO-5 – An oblique surface in an isometric drawing.

Circles cannot be transferred directly to the isometric drawing. As the object is rotated in order to view it as an isometric, holes and cylindrical features also rotate and do not appear as true circles, but as ellipses as in **Figure ISO-6**.

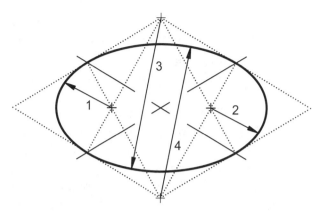

Figure ISO-7 – The four arcs necessary to draw an isometric ellipse. The parallelogram is necessary to properly orient the ellipse and provide starting and ending points for the arcs.

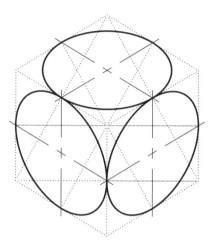

Figure ISO-6 – The parallelogram that is used to inscribe the ellipse changes orientation as it is shifted between the faces of the isometric cube.

The steps in drawing a simple isometric ellipse are illustrated in **Figure ISO-7**:

Step 1 - Draw a parallelogram to scale that frames the completed ellipse. Notice that the orientation of the parallelogram will change based on the orientation of the isometric face.

Step 2 - Find the mid-point of all four sides of the parallelogram by measurement or by drawing a hidden line top to bottom and side to side and mark these intersections.

Step 3 - Draw the four diagonals from the four corners to the mid-points of the opposite sides. The point where these diagonals cross is the radius point for the short radius. The two corners closest together become the radius points for the longer radii.

ISOMETRIC DRAWING OF AN OBJECT WITH CURVED SURFACES

An object with a rounded end, a cylinder and a hole is shown in **Figure ISO-8** The layout is similar to starting an isometric for a rectangular prism. Begin by drawing the framework for the isometric using rectangular shapes. Using this framework, draw the parallelograms so the ellipses are properly oriented. These steps are shown in **Figure ISO-9**.

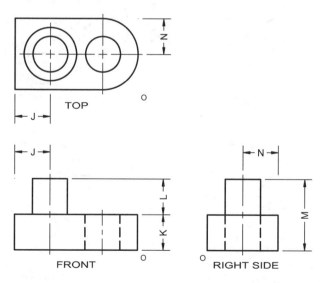

Figure ISO-8 – Orthographic views of an object with curved surfaces.

Step 1 – Sketch the flat plate as a rectangular prism. Using the edges to create a parallelogram, lightly construct the ellipse in both the top and bottom surfaces of the prism.

Step 2 - Draw a light construction line from the center of both ellipses to the opposite corners formed with acute angles. These lines identify where the vertical edge of the rounded end appears. Only half of a parallelogram is needed for the top and bottom surfaces of the rounded end. Construct the parallelogram for the base of the cylinder.

Step 3 – Measuring the distance "L", draw the top parallelogram where the top of the cylinder appears. Draw lines through the centers of the ellipses to the acute angle corners of the parallelograms. These two light construction lines determine where the lower ellipse is visible and the side edges of the cylinder are drawn.

Step 4 – Add the ellipses representing the tops of the two holes.

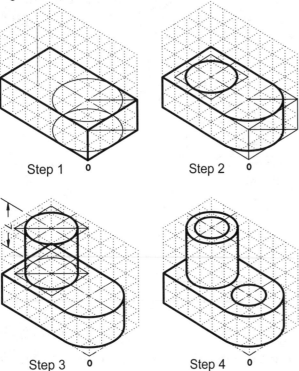

Step 1   0   Step 2   0

Step 3   0   Step 4   0

Figure ISO-9 – Curved surfaces in an isometric.

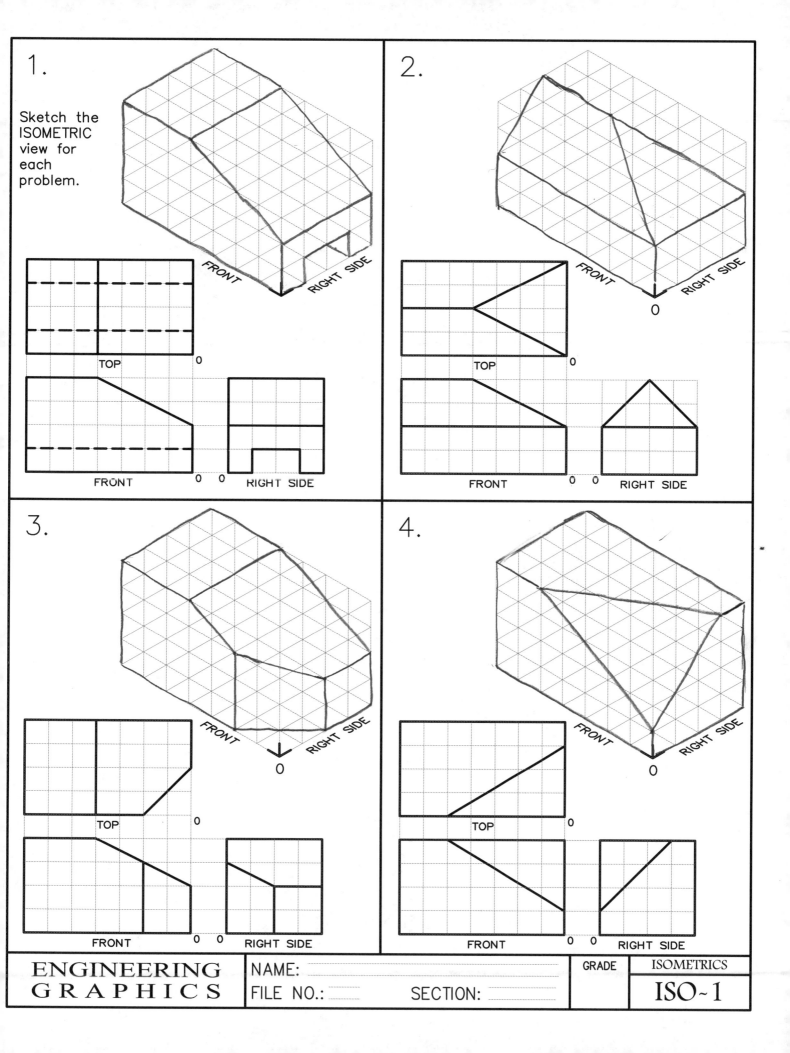

1.

Sketch the ISOMETRIC view for each problem.

FRONT

RIGHT SIDE

TOP

0

FRONT 0 0 RIGHT SIDE

2.

FRONT

RIGHT SIDE

0

TOP

0

FRONT 0 0 RIGHT SIDE

3.

FRONT

RIGHT SIDE

0

TOP

0

FRONT 0 0 RIGHT SIDE

4.

FRONT

RIGHT SIDE

0

TOP

0

FRONT 0 0 RIGHT SIDE

ENGINEERING GRAPHICS

NAME:

FILE NO.:

SECTION:

GRADE

ISOMETRICS

ISO~1

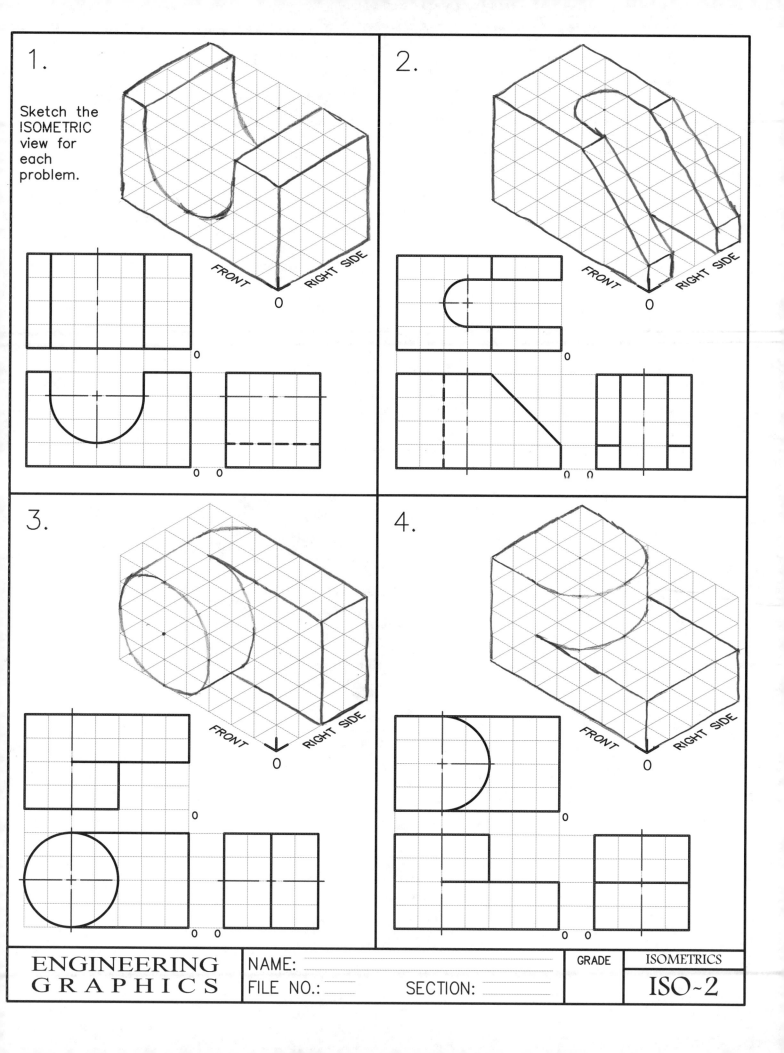

1.

Sketch the
ISOMETRIC
view for
each
problem.

FRONT   RIGHT SIDE
0

2.

FRONT   RIGHT SIDE
0

3.

FRONT   RIGHT SIDE
0

4.

FRONT   RIGHT SIDE
0

ENGINEERING
GRAPHICS

NAME:
FILE NO.:   SECTION:

GRADE   ISOMETRICS

ISO~2

Sketch the ISOMETRIC view of the given object in each problem. Use point "O" for orientation of the ORTHOGRAPHIC views to the ISOMETRIC grid.

1.

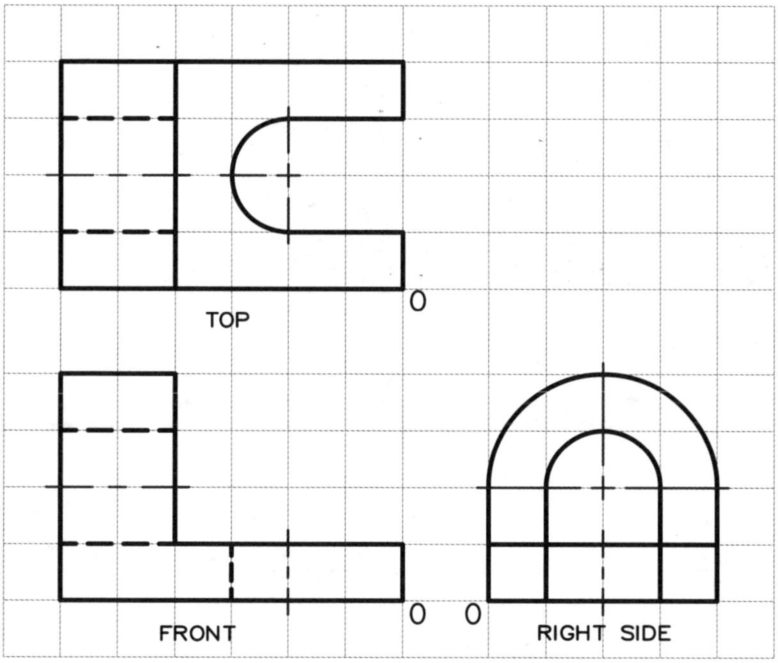

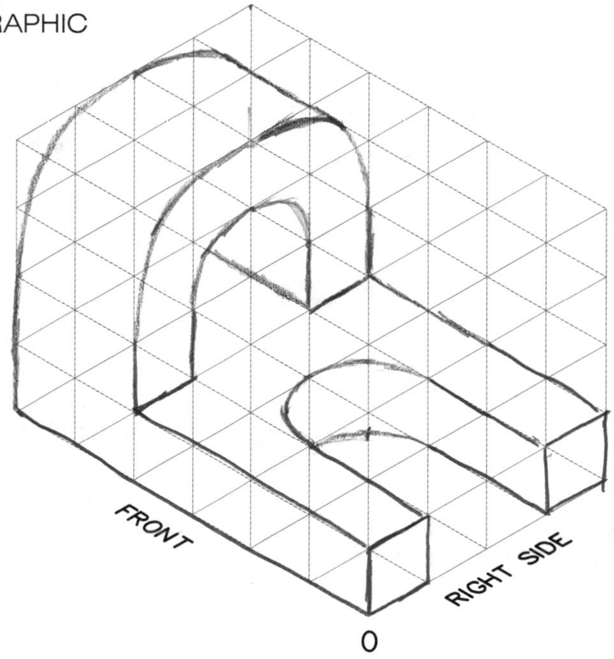

TOP

O

FRONT

O

O

RIGHT SIDE

FRONT

RIGHT SIDE

O

2.

TOP

O

FRONT

O

O

RIGHT SIDE

FRONT

RIGHT SIDE

O

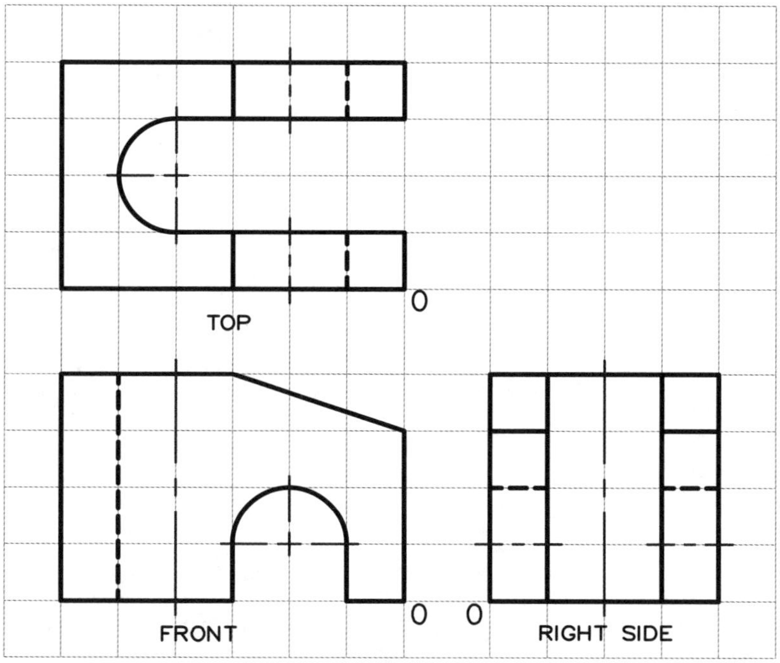

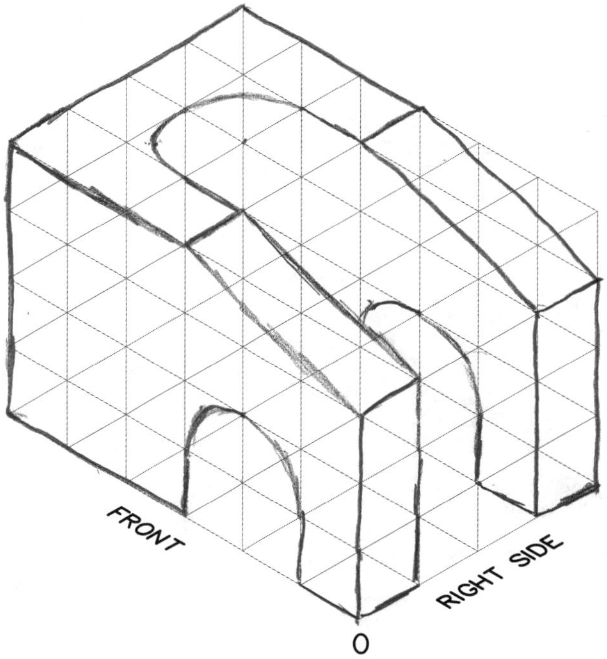

ENGINEERING
GRAPHICS

NAME:

FILE NO.:   SECTION:

GRADE

ISOMETRIC

ISO~3

1.

Sketch the ISOMETRIC view for each problem. Orient the ISOMETRIC view based on the given lines in the ISOMETRIC grid.

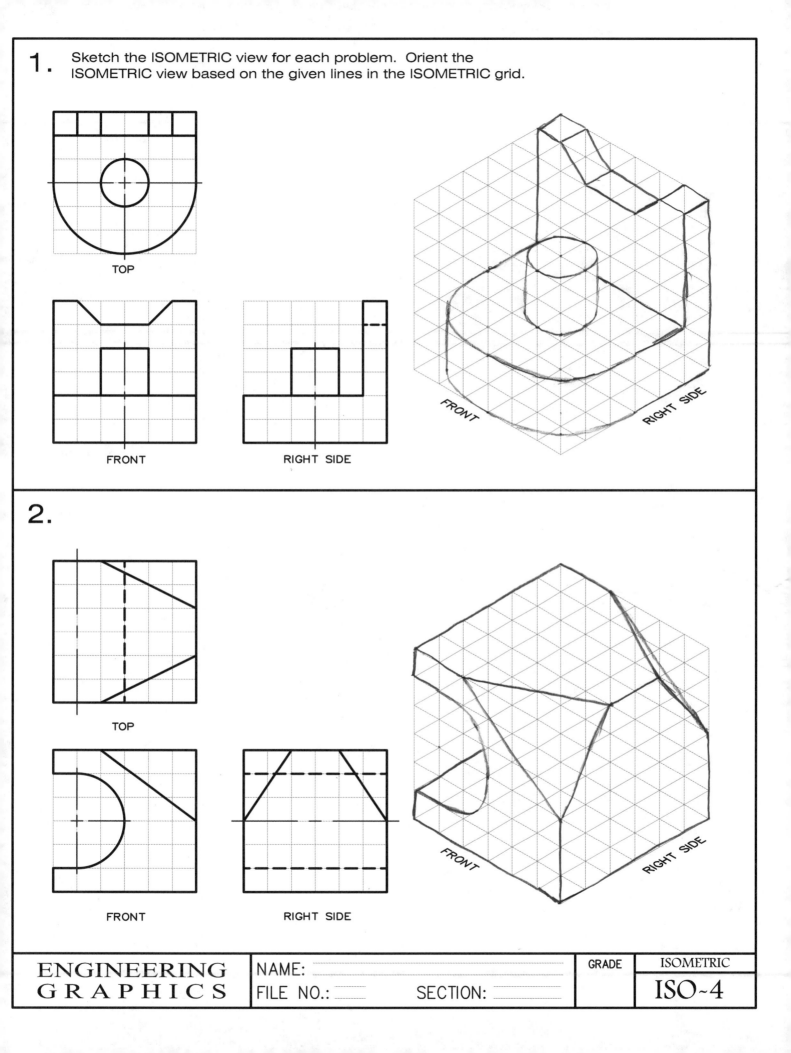

TOP

FRONT

RIGHT SIDE

FRONT

RIGHT SIDE

2.

TOP

FRONT

RIGHT SIDE

FRONT

RIGHT SIDE

ENGINEERING GRAPHICS

NAME:

FILE NO.:   SECTION:

GRADE

ISOMETRIC

ISO-4

1. Shetch the ISOMETRIC view for each problem. Orient the ISOMETRIC view based on the given lines in the ISOMETRIC grid.

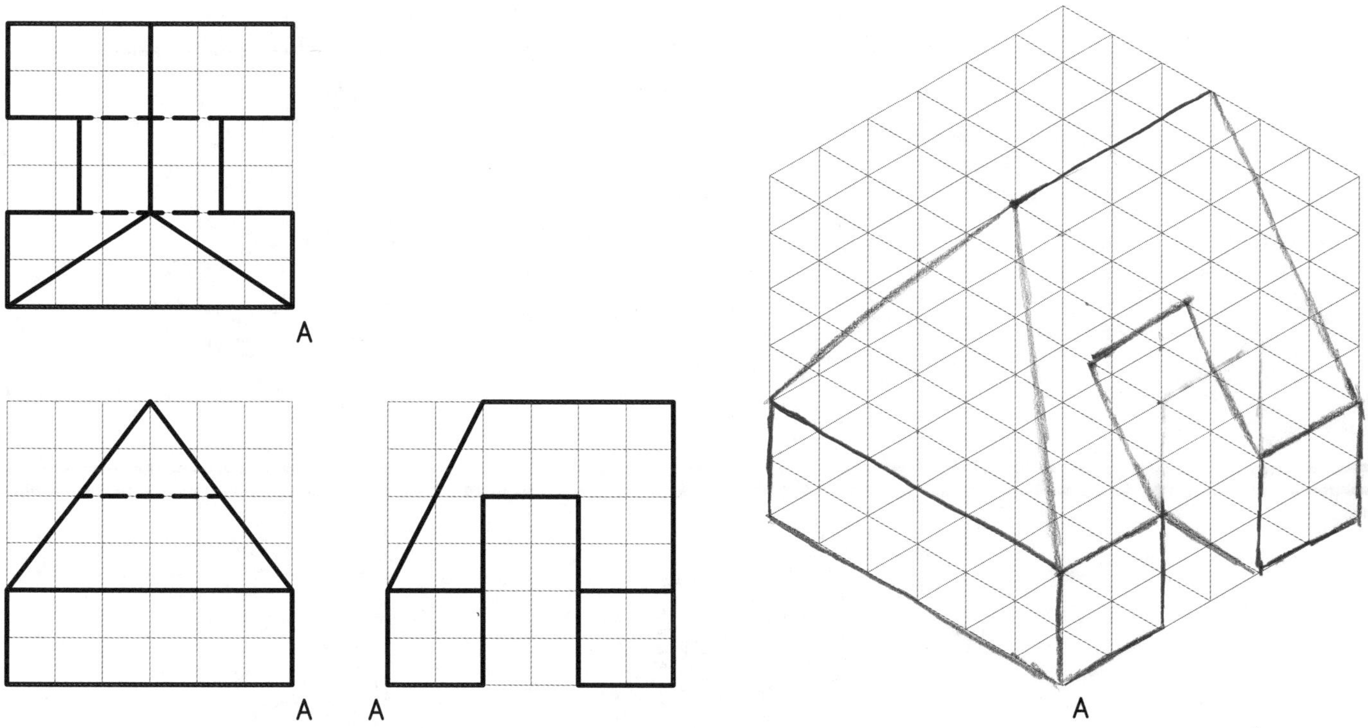

A

A   A

A

2.

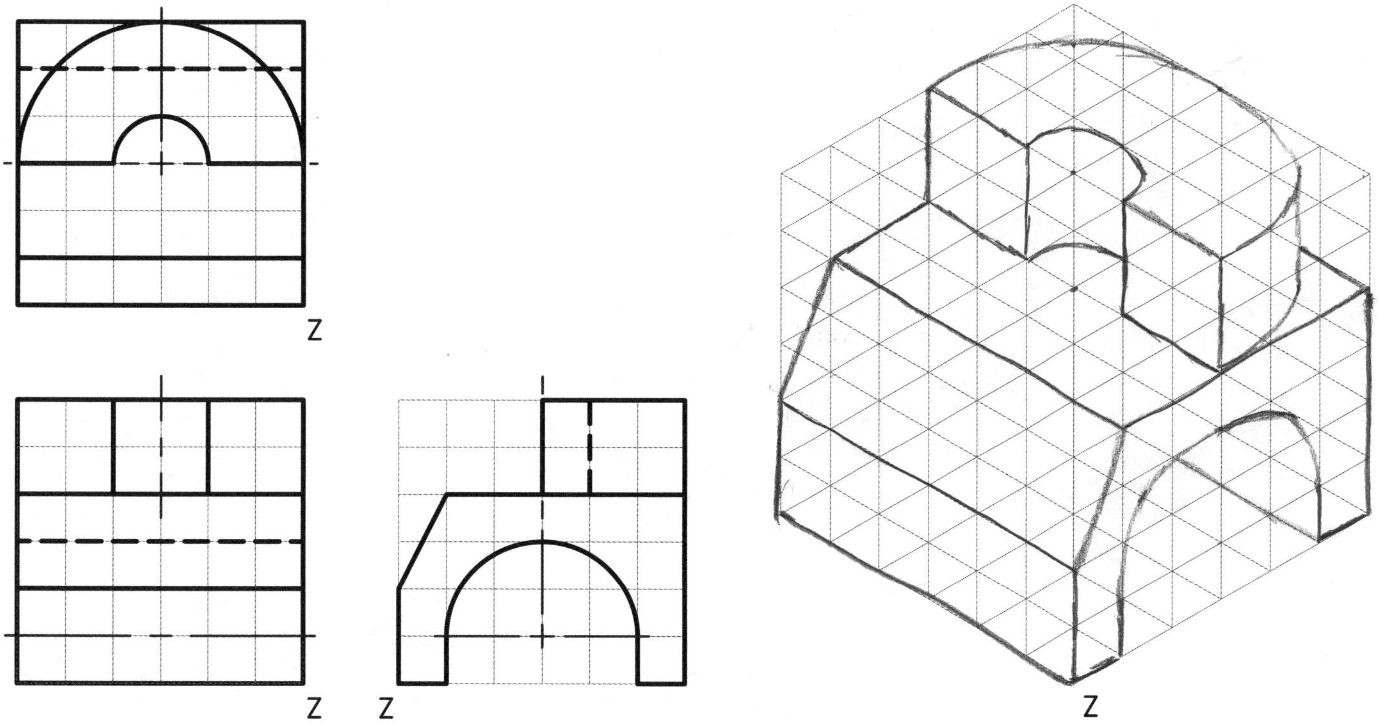

Z

Z   Z

Z

ENGINEERING GRAPHICS

NAME:

FILE NO.:   SECTION:

GRADE

ISOMETRIC

ISO~5

Sketch the ISOMETRIC view of the given object in each problem. Use point "O" for orientation of the ORTHOGRAPHIC views to the ISOMETRIC grid.

1.

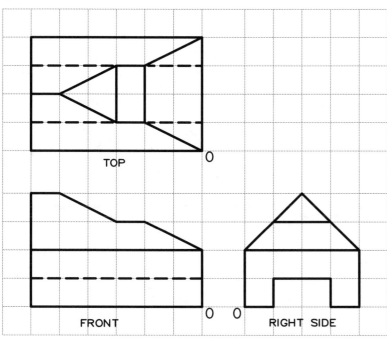

TOP

O

FRONT

O   O

RIGHT SIDE

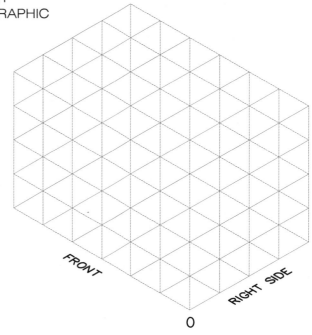

FRONT

RIGHT SIDE

O

2.

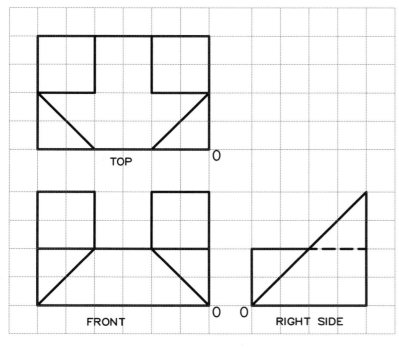

TOP

O

FRONT

O   O

RIGHT SIDE

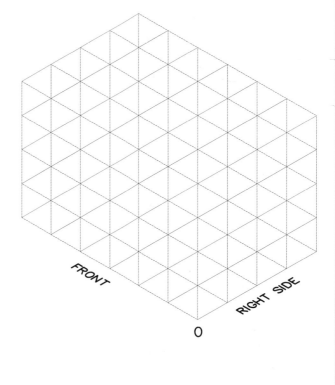

FRONT

RIGHT SIDE

O

| ENGINEERING GRAPHICS | NAME: FILE NO.: SECTION: | GRADE | ISOMETRIC ISO~6 |

Sketch the ISOMETRIC view of the object. The given TOP, FRONT and RIGHT SIDE views are complete. Use point "O" for orientation.

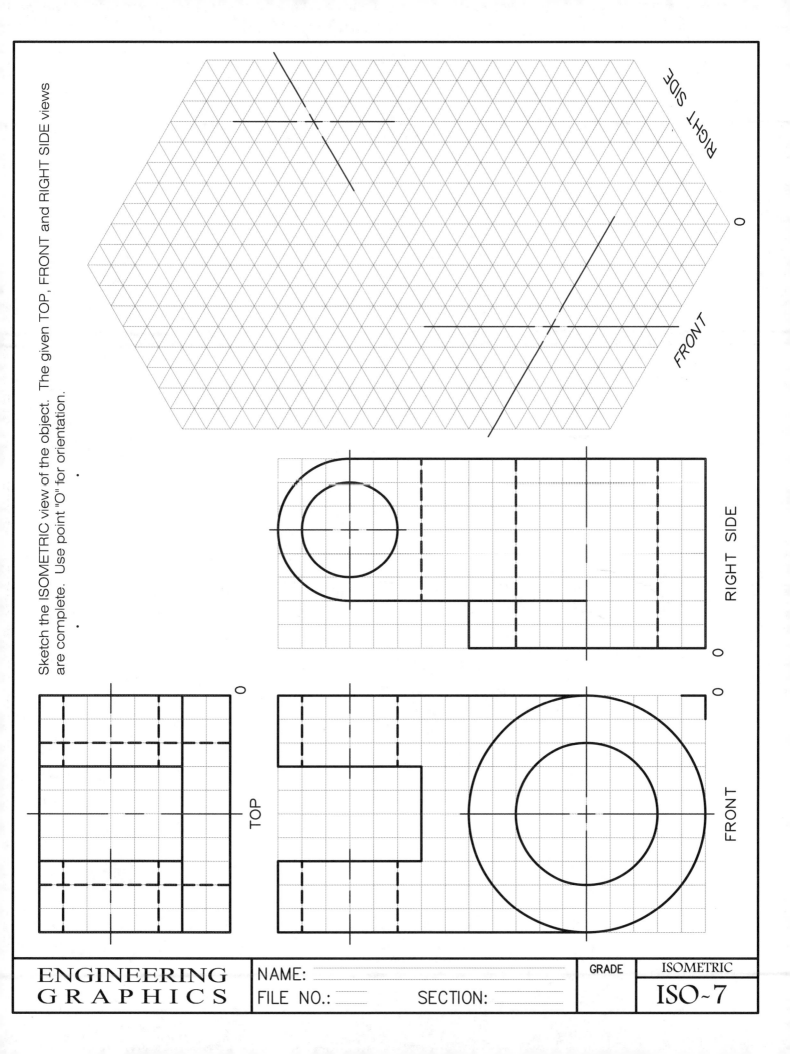

RIGHT SIDE

O

FRONT

RIGHT SIDE

O

TOP

O

FRONT

O

ENGINEERING
GRAPHICS

NAME: _____

FILE NO.: _____   SECTION: _____

GRADE

ISOMETRIC

ISO~7

1.

2.

3.

4.

5.

6.

① 0

② 0

③ 0

④ 0

⑤ 0

⑥ 0

ENGINEERING
GRAPHICS

NAME:

FILE NO.:   SECTION:

GRADE

ISOMETRIC

ISO~8

Sketch the ISOMETRIC view for each problem.

1.

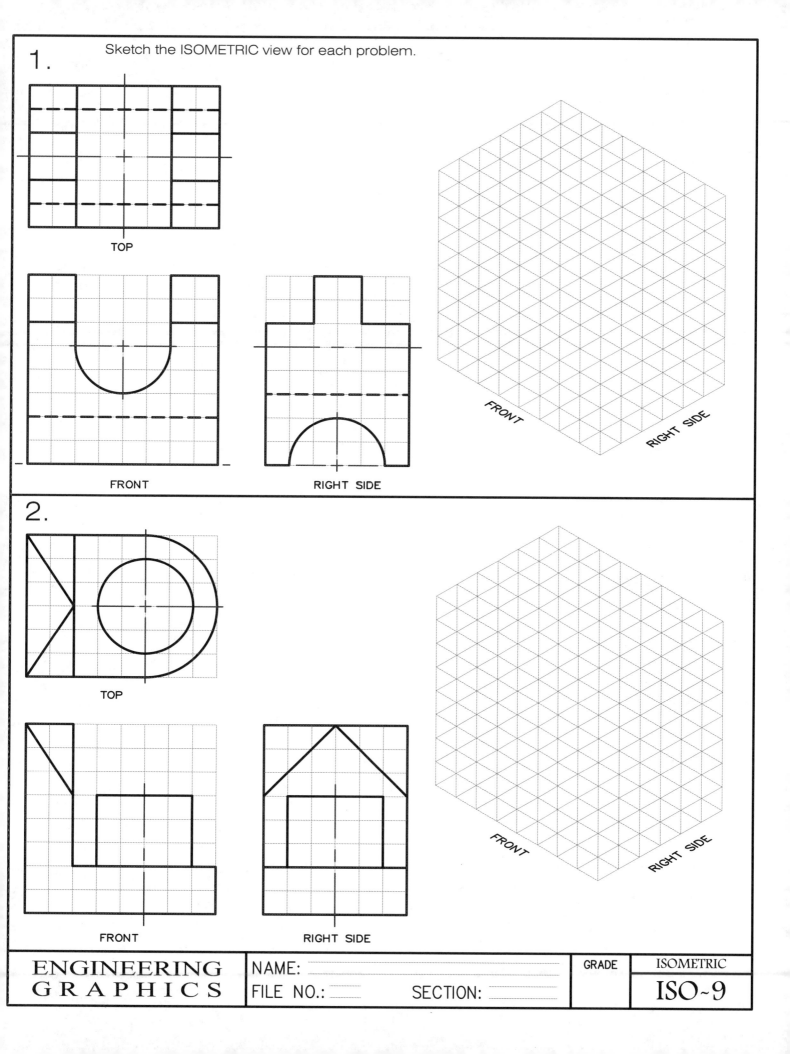

TOP

FRONT

RIGHT SIDE

FRONT

RIGHT SIDE

2.

TOP

FRONT

RIGHT SIDE

FRONT

RIGHT SIDE

| ENGINEERING GRAPHICS | NAME: _____ FILE NO.: _____ SECTION: _____ | GRADE | ISOMETRIC ISO~9 |

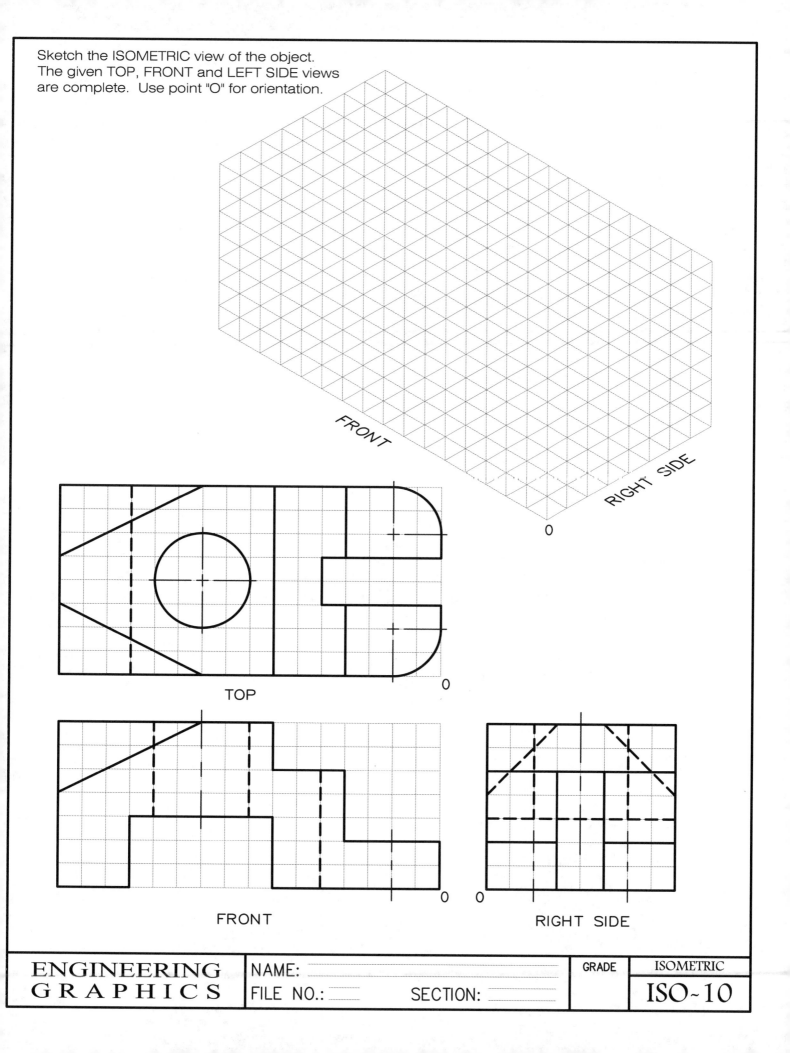

Sketch the ISOMETRIC view of the object.
The given TOP, FRONT and LEFT SIDE views
are complete. Use point "O" for orientation.

FRONT

RIGHT SIDE

0

TOP

0

0

FRONT

0   0

RIGHT SIDE

ENGINEERING
GRAPHICS

NAME:

FILE NO.:   SECTION:

GRADE   ISOMETRIC

ISO-10

Sketch the ISOMETRIC view of the object.
The given TOP, FRONT and LEFT SIDE views
are complete. Use point "O" for orientation.

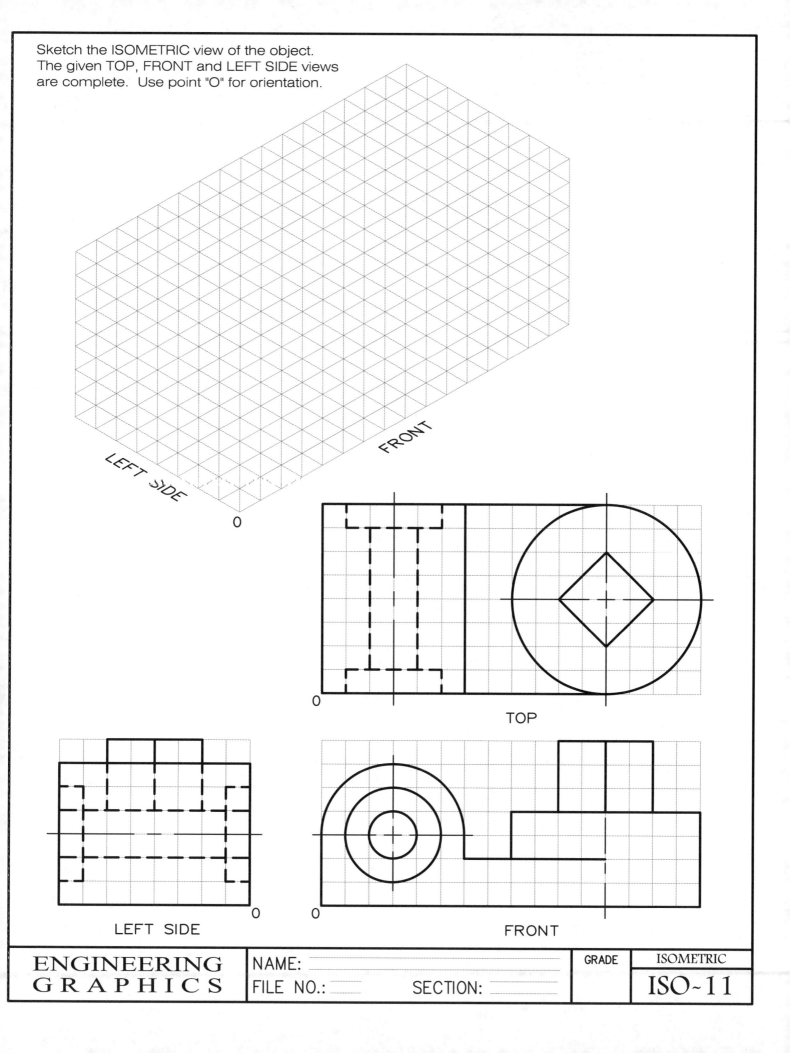

FRONT

LEFT SIDE

0

TOP

0

LEFT SIDE

0

0

FRONT

ENGINEERING
GRAPHICS

NAME:

FILE NO.:   SECTION:

GRADE

ISOMETRIC

ISO~11

OBLIQUE DRAWINGS

Oblique drawings provide a quick way to sketch an object and represent the three dimensions of *height, width* and *depth*. While an isometric drawing creates a more realistic pictorial drawing, the oblique drawing is simple, fast and usually avoids the need for elliptical shapes. Oblique drawings have only one receding axis. This axis can be drawn at any angle, but is typically drawn at a 30° or 45° angle. With a grid system, the 45° receding axis is drawn with a light diagonal line from corner to opposite corner of the grid.

CAVALIER OBLIQUE DRAWINGS

Cavalier oblique drawings transfer the depth measurement from the orthographic projection directly to the receding axis in the oblique drawing. This sometimes makes the oblique drawing appear stretched on the receding axis. However, if the object is simple and has all of its curved surfaces in the same plane, the *cavalier oblique* provides a fast pictorial drawing to represent the object.

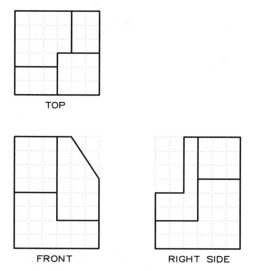

Figure OBL-1 – The orthographic views of the object with rectangular prisms.

Figure OBL-2 shows the steps necessary to draw the Cavalier Oblique of the object in **Figure OBL-1**.

Step 1 - The overall rectangular shape is drawn with the depth measurement placed along the receding axis.

Step 2 - The rectangle or notch on the right hand side is drawn to show where the corner has been removed.

Step 3 - The rectangular notch on the upper left front corner is drawn.

Step 4 - The inclined surface on the upper right rear edge is located by marking the corners of the inclined surface and then connecting those corners to generate the angle. As you construct the oblique drawing, notice that surfaces that are parallel in the orthographic view remain parallel when transferred to the oblique drawing

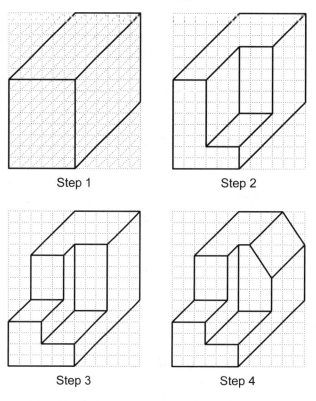

Figure OBL-2 – The steps in drawing a cavalier oblique. Notice in Step 1 that the receding axis in shown and goes diagonally from corner to corner. The receding axis is removed in the following steps for clarity.

CABINET OBLIQUE DRAWINGS

A *cabinet oblique* drawing is constructed similar to a cavalier oblique drawing except the distances transferred along the receding axis are reduced by half. This reduces the exaggerated (or long) look of oblique drawings where the depth measurement is significantly greater in proportion to the height and width.

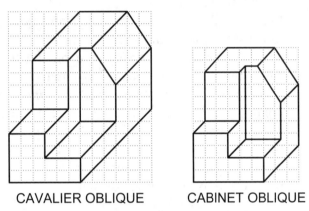

CAVALIER OBLIQUE   CABINET OBLIQUE

Figure OBL-3 – The difference between a Cavalier and Cabinet Oblique drawing is readily apparent in this comparison. The shortened receding axis reduces the exaggerated length of the depth measurement.

CURVED SURFACES IN OBLIQUE DRAWINGS

And oblique drawling lends its self to quick representations of objects with curved surfaces. When the curved edges of cylindrical features and holes appear circular in the front face of the object, the radii and diameters can be transferred directly to the appropriate face in the oblique drawing. The surface that appears curved in the orthographic projection will also appear as the same curved surface in the oblique drawing. This eliminates the need for elliptical shapes and speeds up the drawing process. To determine if the back edge of a hole will appear in the oblique drawing, simply move the center point back along the receding axis the appropriate distance. Move the circle back to this second center point and determine if the edge is visible.

Figure OBL-5 shows the steps necessary to draw the cavalier oblique of the object depicted in **Figure OBL-4**. The steps to construct the oblique with curved surfaces are the same as the steps in constructing a rectangular prism.

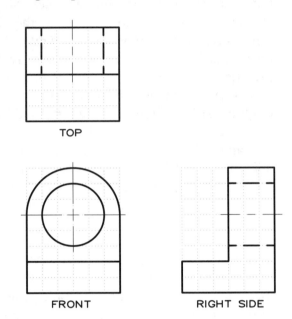

TOP

FRONT   RIGHT SIDE

Figure OBL-4 – The orthographic views of the object with curved surfaces.

Step 1 - Begin the oblique drawing with light construction lines for the rectangular base plate and rear vertical plate.

Step 2 - The two circles representing the curved top of the vertical rear plate are drawn in both the front face and rear face.

Step 3 - The light diagonal line that goes from the center of the circle to the upper left corner in each face defines the point of tangency for the line that represents the edge of the curved surface on the upper left side. This diagonal line determines where the curved surface on the rear face ends and the line representing the edge begins. These two diagonal lines are perpendicular to the receding axes. The circle representing the hole in the vertical plate is drawn in the front surface with an object line and in the rear surface with a light construction line to check for visibility.

Step 4 - Since the circles representing the hole do not overlap, you will not see the edge of the rear hole. Therefore, only the circle representing the front edge of the hole is drawn with a visible line.

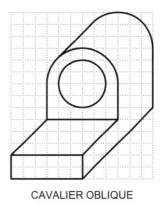
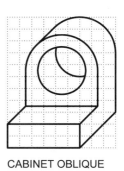

CAVALIER OBLIQUE   CABINET OBLIQUE

Figure OBL-6 – With the shorter receding axis, the rear edge of the hole appears in the Cabinet oblique.

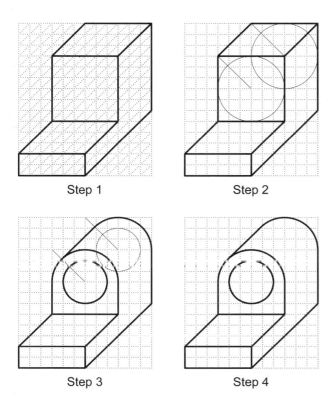

Step 1   Step 2

Step 3   Step 4

Figure OBL-5 – The steps in drawing the cavalier oblique of an object with curved surfaces. Notice in Step 1 that the receding axis in shown and goes diagonally from corner to corner. The receding axis in the following steps is removed for clarity.

Figure OBL-6 shows the object with curved surfaces as it would appear in both a cavalier and a cabinet oblique drawing. Although the receding axis is 45° in both drawings, the cabinet oblique looks less exaggerated along the receding axis. Notice that the rear edge of the hole does appear in the cabinet oblique since the shorter receding axis brings the edge of the hole forward.

1.

Sketch the OBLIQUE drawings based on the ORTHOGRAPHIC views.

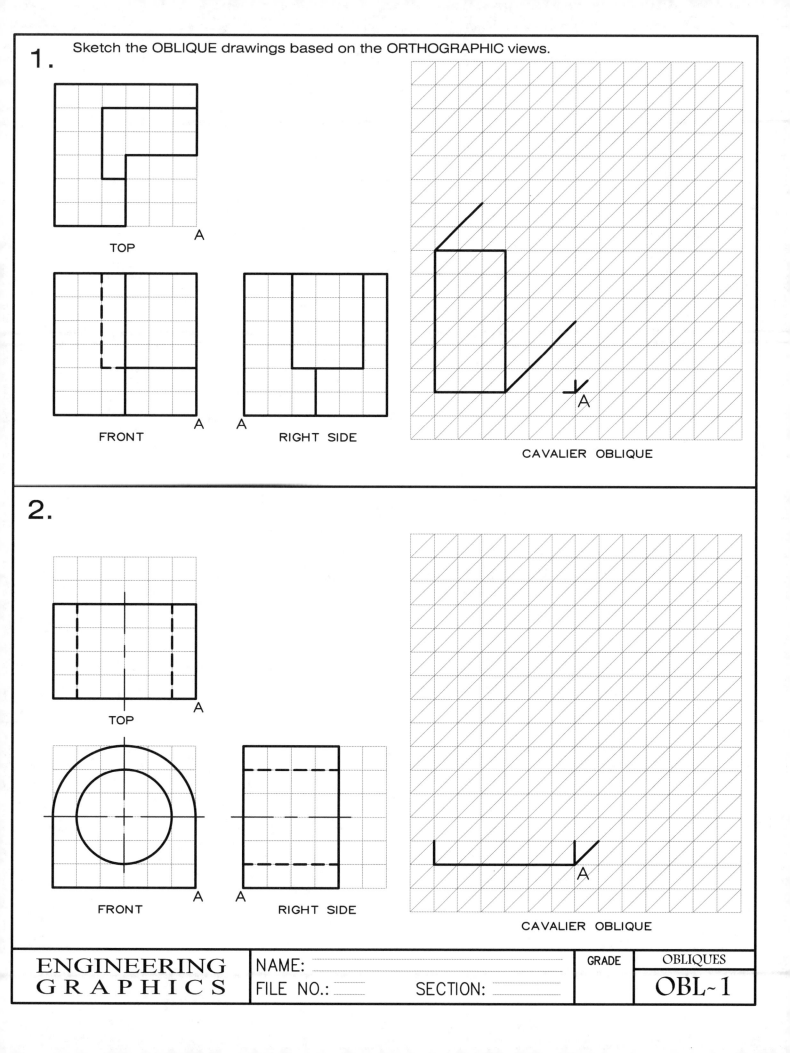

TOP

FRONT

RIGHT SIDE

CAVALIER OBLIQUE

2.

TOP

FRONT

RIGHT SIDE

CAVALIER OBLIQUE

ENGINEERING
GRAPHICS

NAME:

FILE NO.:   SECTION:

GRADE

OBLIQUES

OBL~1

1.

Sketch the OBLIQUE drawings based on the ORTHOGRAPHIC views.

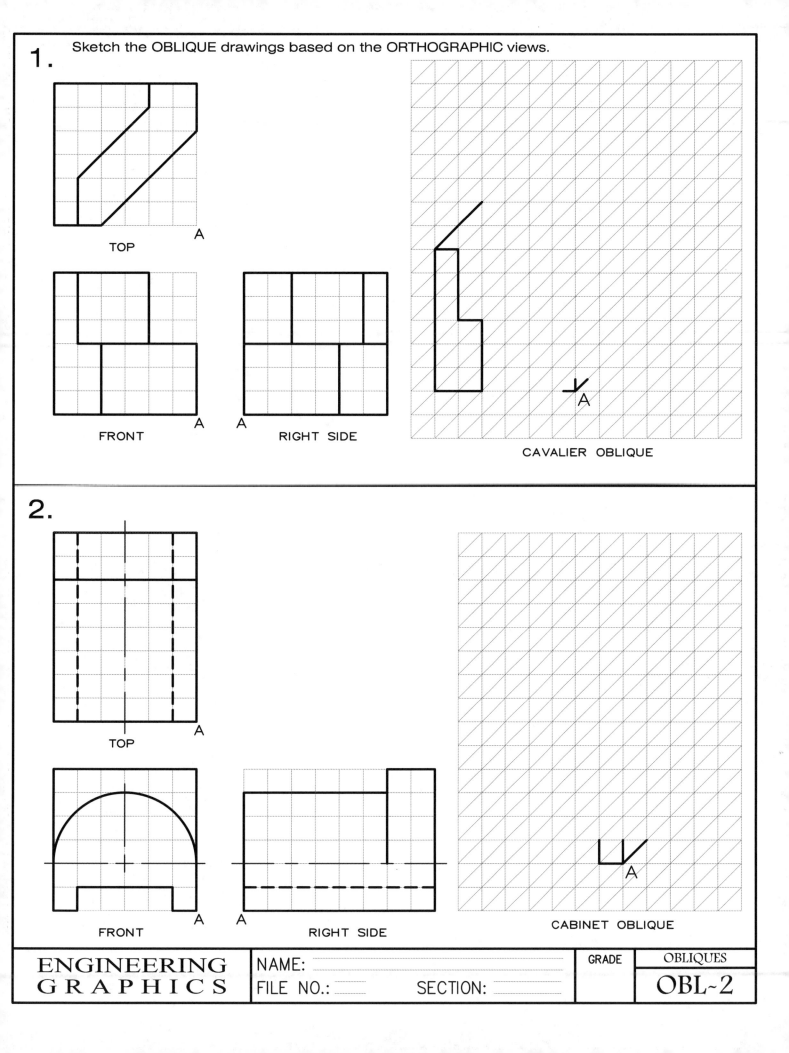

TOP

FRONT

RIGHT SIDE

CAVALIER OBLIQUE

2.

TOP

FRONT

RIGHT SIDE

CABINET OBLIQUE

ENGINEERING
GRAPHICS

NAME:

FILE NO.:   SECTION:

GRADE   OBLIQUES

OBL~2

Sketch the OBLIQUE drawings based on the ORTHOGRAPHIC views.

1.

TOP

FRONT

RIGHT SIDE

CAVALIER OBLIQUE

2.

TOP

FRONT

RIGHT SIDE

CABINET OBLIQUE

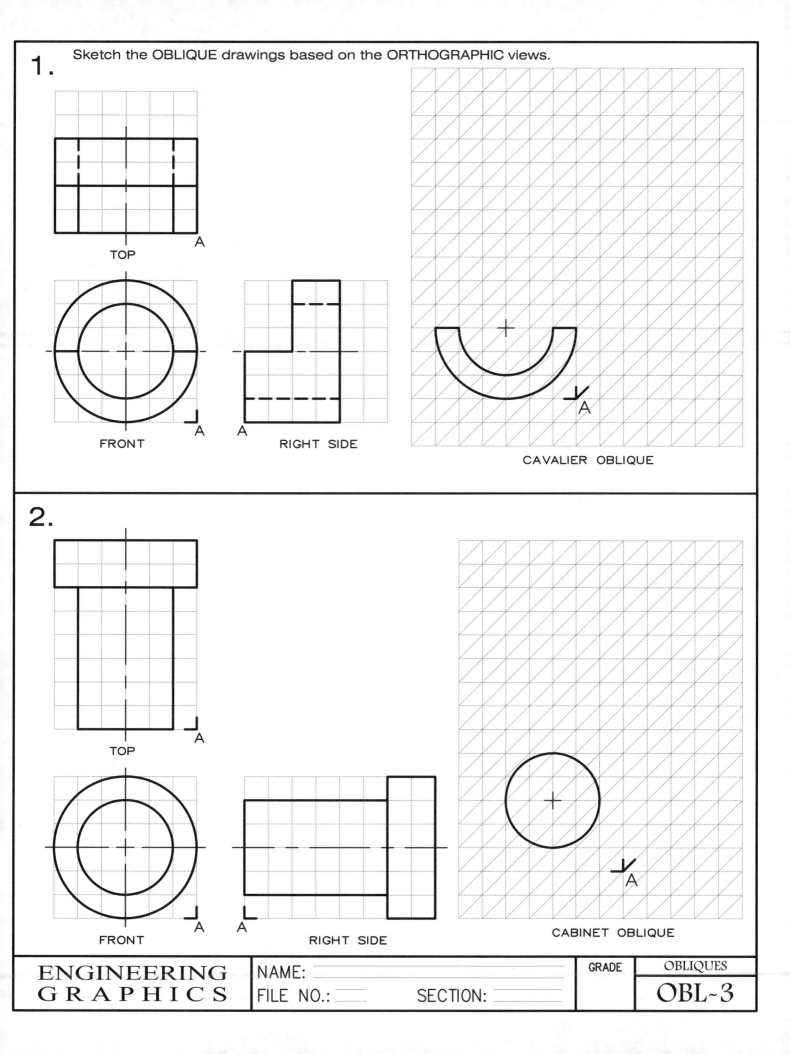

ENGINEERING GRAPHICS

NAME:

FILE NO.:   SECTION:

GRADE

OBLIQUES

OBL~3

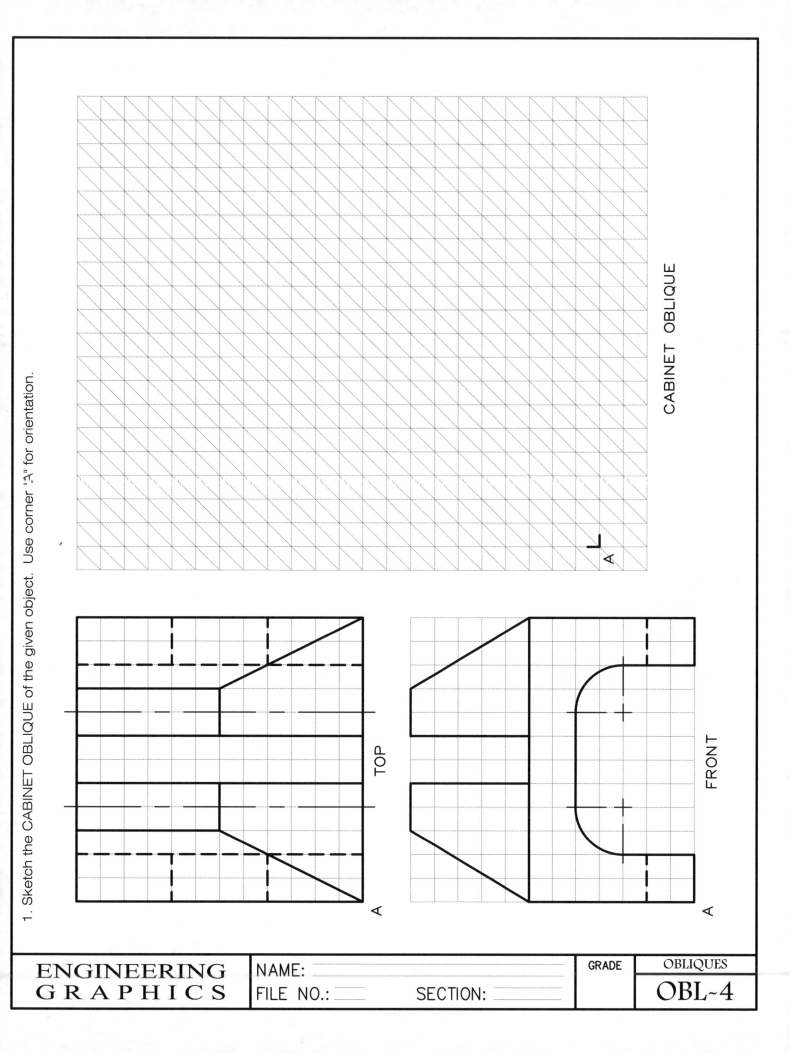

1. Sketch the CABINET OBLIQUE of the given object. Use corner "A" for orientation.

CABINET OBLIQUE

A

TOP

A

FRONT

A

ENGINEERING
GRAPHICS

NAME: _____

FILE NO.: _____ SECTION: _____

GRADE

OBLIQUES

OBL~4

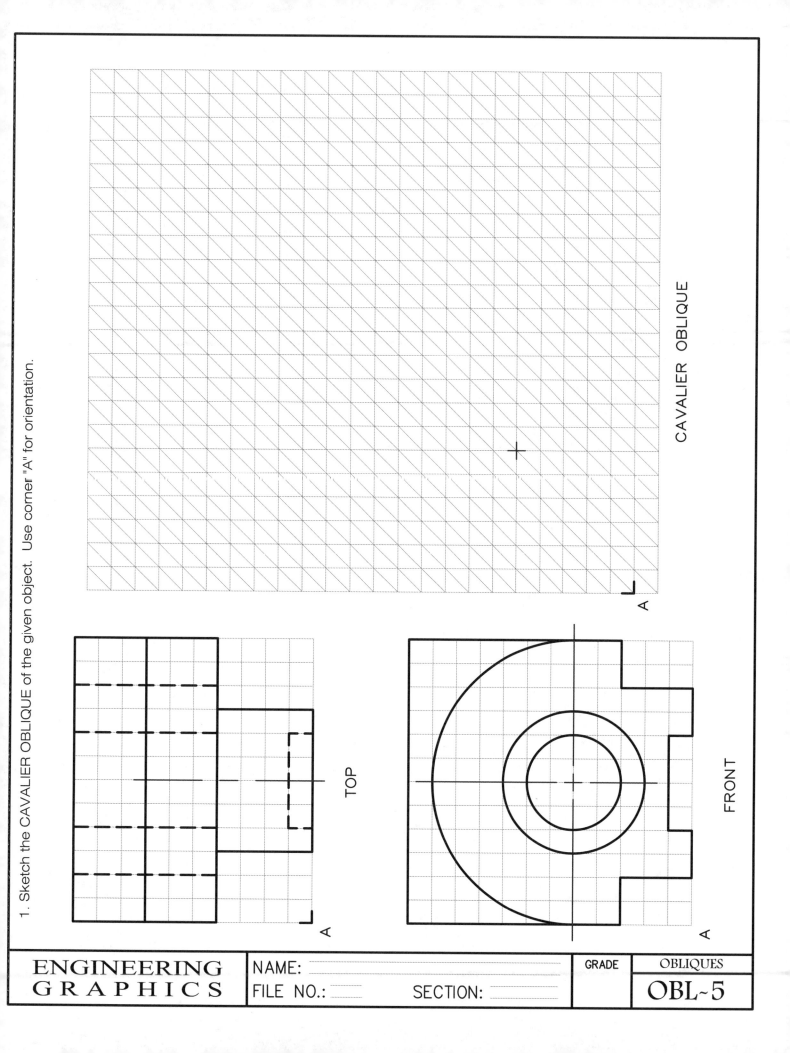

1. Sketch the CAVALIER OBLIQUE of the given object. Use corner "A" for orientation.

CAVALIER OBLIQUE

A

TOP

A

FRONT

A

ENGINEERING
GRAPHICS

NAME:

FILE NO.:   SECTION:

GRADE

OBLIQUES

OBL~5

Sketch the CABINET OBLIQUE of the given object. Use corner "A" for orientation.

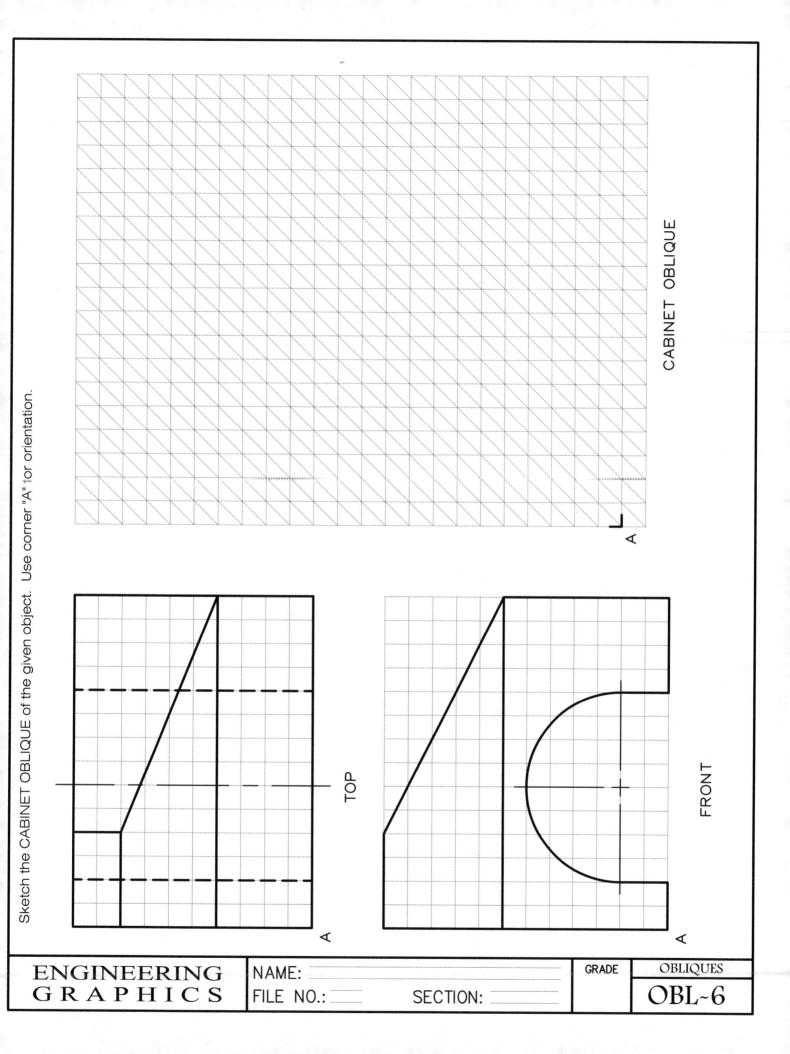

CABINET OBLIQUE

TOP

A

FRONT

A

ENGINEERING
GRAPHICS

NAME: _____

FILE NO.: _____   SECTION: _____

GRADE

OBLIQUES

OBL~6

AUXILIARY VIEWS

For most objects, two or three views drawn orthographically are sufficient to describe the object. However some objects, especially those having an inclined surface, require an additional view to show the inclined surface true size and shape. An auxiliary view is used to provide this true size and shape view. The inclined surface in **Figure AUX-1** appears as an edge view in the front view.

in the front view) with the true length depth measurement from either the top or right side view and creates a true size and shape view of the inclined surface.

A fold line is drawn between views. The distance from the fold line to the front edge of the top view and from the fold line to the front edge of the right side view will be the same measurement. These fold lines are perpendicular to one another and parallel to the edge of the views. Another fold line is drawn between the front view and the auxiliary view. This fold line is drawn parallel to the edge view of the inclined surface in the front view.

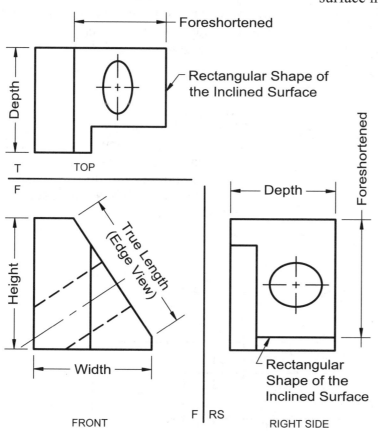

Figure AUX-1 – The inclined surface appears as an edge in the front view. The surface is foreshortened in the top and right side views.

In the top and right side views, this inclined surface appears rectangular but does not appear true size and shape.

The auxiliary view incorporates the true length measurement of the edge view (found

When drawing orthographically, projections between views are made perpendicular to the fold lines between views. This technique of projecting perpendicular to the fold line is continued when projecting measurements from the front view to the auxiliary view. The true length measurement indicated in the front view is projected into the auxiliary view with light construction lines. The necessary depth measurement is transferred from the top view or right side view. Begin by transferring the distance back from the fold line to the leading edge of the auxiliary view. Continue by transferring the remaining depth measurements.

Circular features in the inclined surface will appear elliptical in the rectangular views of the inclined surface. However, these circular features will appear as complete circles or arcs in the auxiliary view since the direction of sight is perpendicular to the surface.

Typically, only the true size and shape of the inclined surface is drawn in the auxiliary view. If it is preferable to draw the entire object in the auxiliary view, the remaining points of the object are projected.

AUXILIARY VIEW OF AN INCLINED SURFACE

The steps in drawing an Auxiliary View are presented in separate illustrations.

Step One (Figure AUX-2) - The corners of the inclined plane are numbered in the right side view (where the surface appears foreshortened) for easy reference. (The top view could also be used instead of the right side view.) These points are projected across to the front view (where the inclined surface appears as an edge view) and numbered accordingly.

Figure AUX-2
STEP 1 – The corners are marked in the foreshortened view and projected to the edge view of the surface. Either the right side or top view can be used since both have the depth measurement.

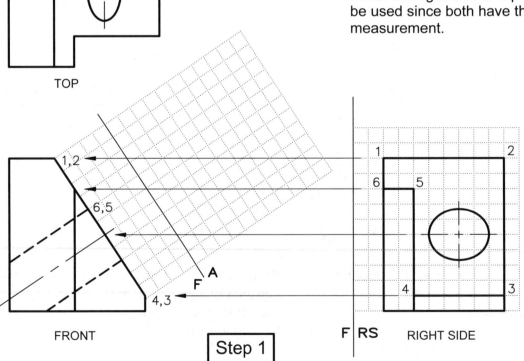

Step Two (Figure AUX-3) - The fold line between the front view and the auxiliary view is added and drawn parallel to the edge view of the inclined surface. This fold line can be located at any distance from the edge view of the inclined surface, but the fold line must be parallel to the edge view. The corners of the inclined surface are projected from the front view, perpendicular to the fold line (between the front view and the auxiliary view), and located in the auxiliary view. Notice that the distance from the fold line to the leading edge of the object (measurement "A") is the same in both the right side view and the auxiliary view. The distances from the fold line to the remaining points are transferred and measured from the fold line in the auxiliary view.

Figure AUX-3
STEP 2 – The corners are projected into the auxiliary view.

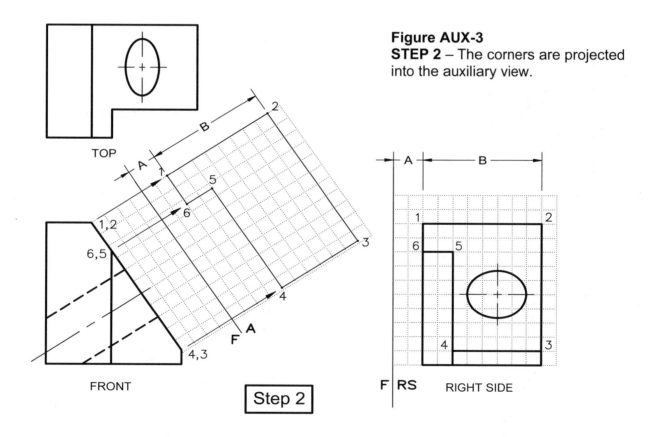

Step Three (Figure AUX-4) - The points are connected to draw the outline of the inclined surface that appears true size and shape in the auxiliary view. The hole, which appears elliptical in the top and right side views, will appear as a true circle in the auxiliary view because the angle of sight will be directly down the centerline of the hole. The depth measurement in the top and right side views is the true length measurement of the diameter of the hole. The distance between the hidden lines in the front view also indicates the true diameter of the hole. The center mark is located by projecting the centerline and using the measurement from the fold line to the center mark (measurement "C"). The diameter of the circle is also projected from the front view and corresponds to measurement "D" in the top and right side views.

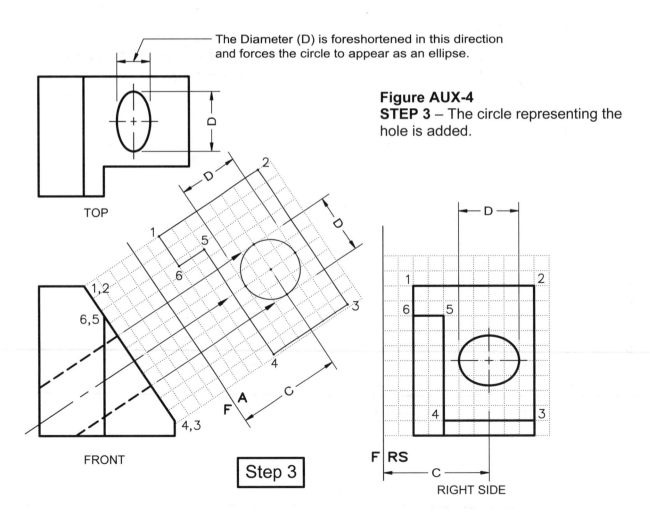

The Diameter (D) is foreshortened in this direction and forces the circle to appear as an ellipse.

Figure AUX-4
STEP 3 – The circle representing the hole is added.

TOP

FRONT

Step 3

F|RS

RIGHT SIDE

Step Four (Figure AUX-5) - The circle representing the hole is drawn and the centerlines for the hole are added. Drawing object lines over construction lines of the surface completes the auxiliary view.

projecting. In this example, the front view shows **height** and **width**. Therefore, the missing dimension of **depth** is needed and can be transferred from the top or right side views.

Figure AUX-5

STEP 4 – The light construction lines are converted to object lines and the surface appears true size and shape in the auxiliary view.

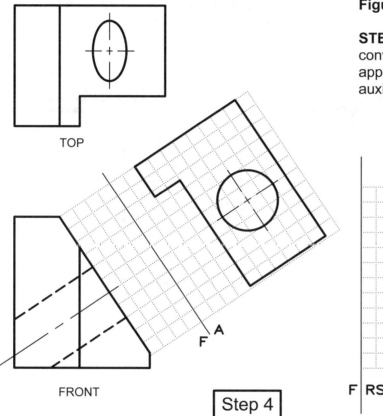

TOP

FRONT

Step 4

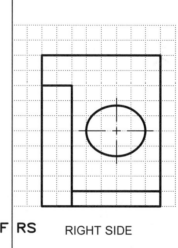

F | RS   RIGHT SIDE

This illustration projects the auxiliary view from the front view. However, an auxiliary view can be projected from any of the remaining primary orthographic views, in any direction, as long as the fold line is parallel to the edge view of the inclined surface.

The technique of numbering the corners is a simple way to ensure that the corners are projected to the correct locations and the appropriate measurements are transferred. Another way to determine the missing dimension necessary for the auxiliary view is to analyze the view from which you are

If projecting the auxiliary view from the top view, project the measurement of the edge view of the inclined surface into the auxiliary view and add one other dimension. Since the top view shows **width** and **depth** the missing dimension needed is **height**.

Projecting an auxiliary view from the right side view requires the transfer of the **width** measurement from either the top or front view because the right side view will include **depth** and **height**. In each of the three examples, the process is to combine the edge view measurement (that appears true length) with the missing measurement projected from an adjoining view.

The entire object can be drawn in the auxiliary view. The remaining corners are projected to the auxiliary view and located.

These corners can be labeled or numbered for easy identification while transferring measurements. Hidden lines will appear to indicate edges behind the visible surfaces.

Figure AUX-6.– The entire object can be shown in the auxiliary view.

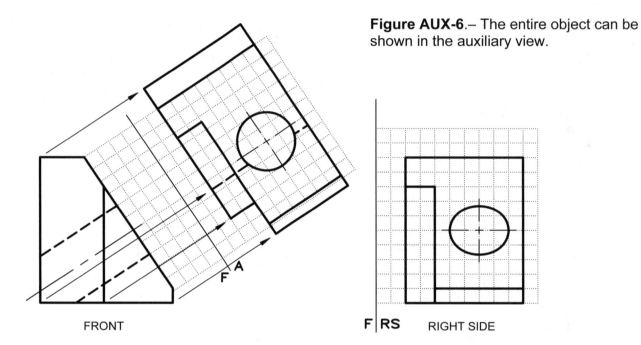

FRONT

F | RS   RIGHT SIDE

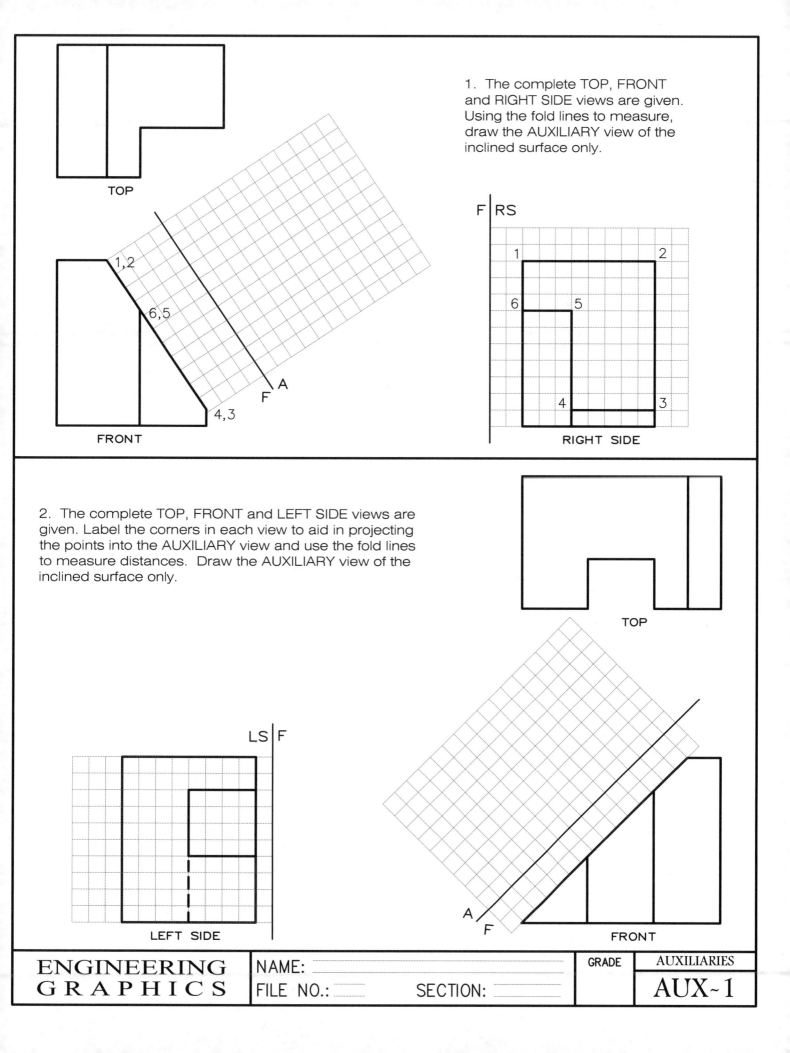

TOP

1. The complete TOP, FRONT and RIGHT SIDE views are given. Using the fold lines to measure, draw the AUXILIARY view of the inclined surface only.

F | RS

1   2
6   5
4   3

RIGHT SIDE

1,2
6,5
4,3
A
F

FRONT

2. The complete TOP, FRONT and LEFT SIDE views are given. Label the corners in each view to aid in projecting the points into the AUXILIARY view and use the fold lines to measure distances. Draw the AUXILIARY view of the inclined surface only.

TOP

LS | F

LEFT SIDE

A
F
FRONT

ENGINEERING GRAPHICS

NAME: _____
FILE NO.: _____   SECTION: _____

GRADE | AUXILIARIES

AUX~1

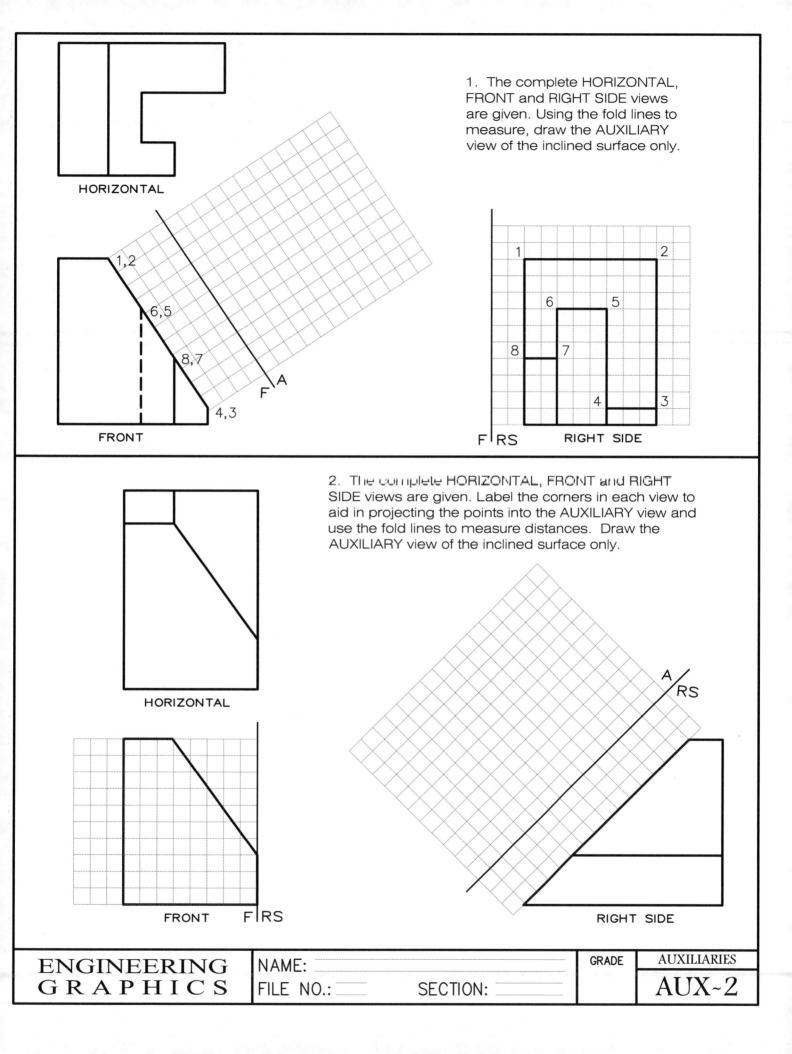

HORIZONTAL

1. The complete HORIZONTAL, FRONT and RIGHT SIDE views are given. Using the fold lines to measure, draw the AUXILIARY view of the inclined surface only.

FRONT

1,2
6,5
8,7
4,3

F A

1   2
6   5
8   7
4   3

F RS   RIGHT SIDE

2. The complete HORIZONTAL, FRONT and RIGHT SIDE views are given. Label the corners in each view to aid in projecting the points into the AUXILIARY view and use the fold lines to measure distances. Draw the AUXILIARY view of the inclined surface only.

HORIZONTAL

FRONT   F RS

A
RS

RIGHT SIDE

ENGINEERING GRAPHICS

NAME:
FILE NO.:   SECTION:

GRADE   AUXILIARIES

AUX~2

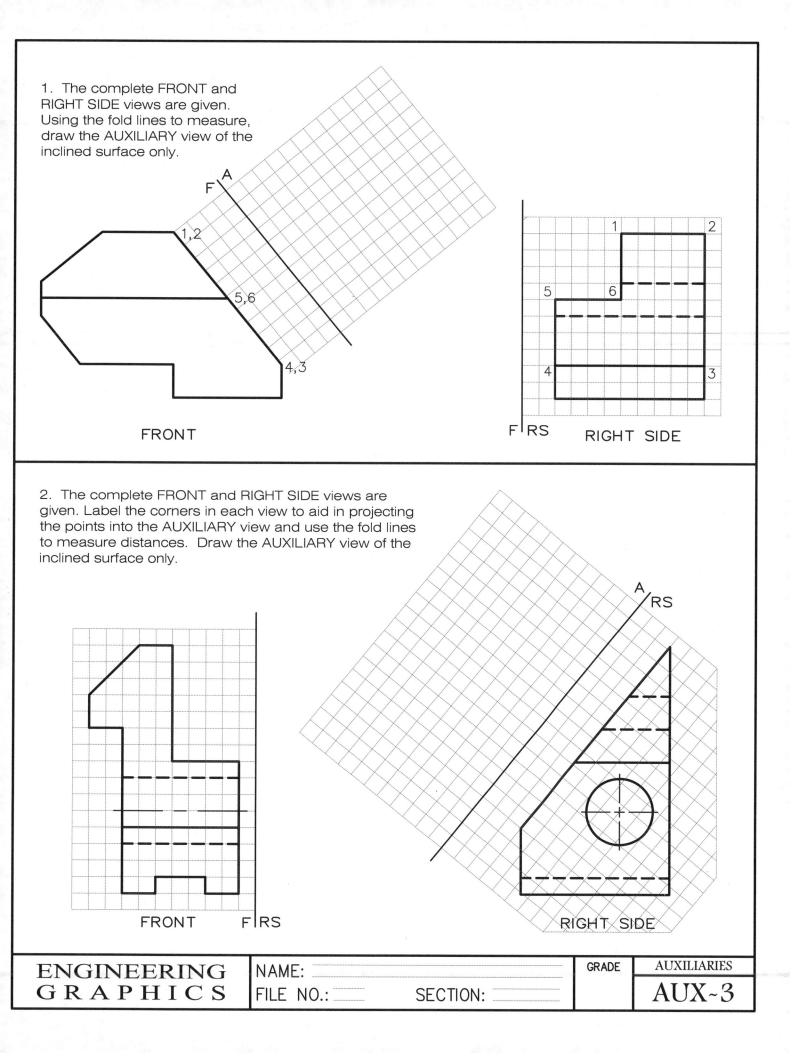

1. The complete FRONT and RIGHT SIDE views are given. Using the fold lines to measure, draw the AUXILIARY view of the inclined surface only.

F A

1,2

5,6

4,3

FRONT

1   2

5   6

4   3

F RS   RIGHT SIDE

2. The complete FRONT and RIGHT SIDE views are given. Label the corners in each view to aid in projecting the points into the AUXILIARY view and use the fold lines to measure distances. Draw the AUXILIARY view of the inclined surface only.

A RS

FRONT   F RS

RIGHT SIDE

ENGINEERING GRAPHICS

NAME: _____

FILE NO.: _____   SECTION: _____

GRADE

AUXILIARIES

AUX~3

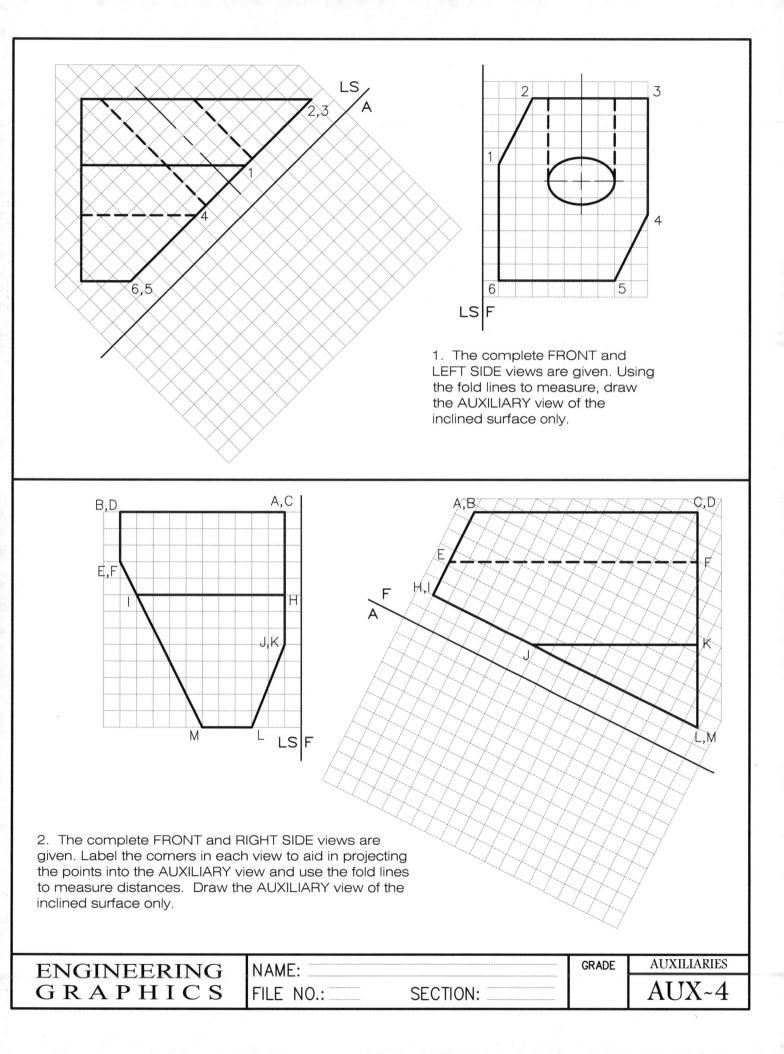

1. The complete FRONT and LEFT SIDE views are given. Using the fold lines to measure, draw the AUXILIARY view of the inclined surface only.

2. The complete FRONT and RIGHT SIDE views are given. Label the corners in each view to aid in projecting the points into the AUXILIARY view and use the fold lines to measure distances. Draw the AUXILIARY view of the inclined surface only.

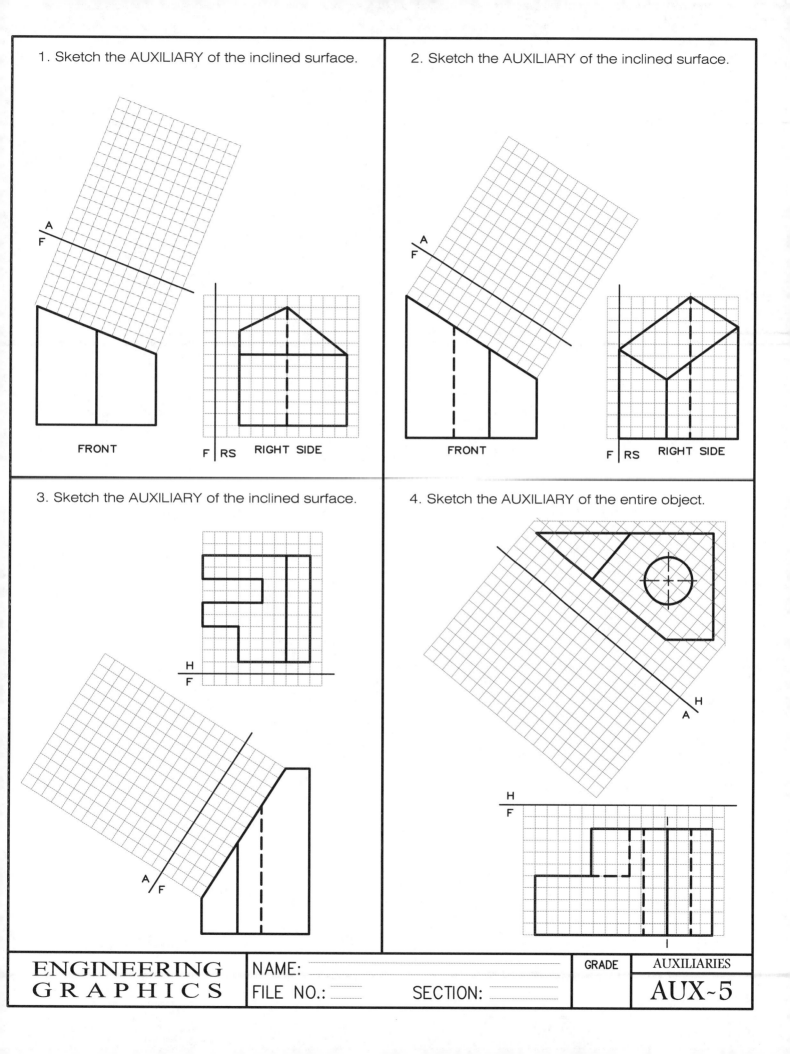

1. Sketch the AUXILIARY of the inclined surface.

A
F

FRONT

F | RS   RIGHT SIDE

2. Sketch the AUXILIARY of the inclined surface.

A
F

FRONT

F | RS   RIGHT SIDE

3. Sketch the AUXILIARY of the inclined surface.

H
F

A / F

4. Sketch the AUXILIARY of the entire object.

H
A

H
F

ENGINEERING
GRAPHICS

NAME: _____
FILE NO.: _____   SECTION: _____

GRADE   AUXILIARIES

AUX~5

1. The complete FRONT and LEFT SIDE views are given. Using the fold lines to measure, draw the AUXILIARY view of the inclined surface. Label corners for reference.

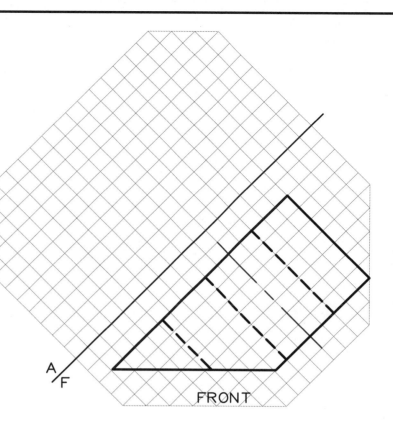

LS F

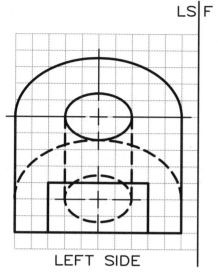

LEFT SIDE

A
F

FRONT

2. The complete FRONT and LEFT SIDE views are given. Using the fold lines to measure, draw the AUXILIARY view of the inclined surface. Label corners for reference.

A
F

F LS

ENGINEERING GRAPHICS	NAME:		GRADE	AUXILIARIES
	FILE NO.:	SECTION:		AUX~6

SECTIONS

The orthographic views are usually sufficient to describe the shape of an object. Sometimes hidden lines make it difficult to visualize complex internal features. A section view clarifies the shape and location of the internal features. In a section view the object is sliced with an imaginary cutting plane. The portion closest to the viewer is removed so that the internal features of the object are visible.

Most section views require a cutting plane line to indicate where the object is divided. This line is typically a phantom line (a series of two short segments and one longer line segment), a hidden line type (with longer dashes all the same length) or end arrows if the cutting plane line would hide important details in the drawing.

FULL SECTIONS

If the cutting plane goes all the way across the object in a straight line, it is termed a *full section*. The section is typically drawn in an adjacent view and replaces that orthographic view. In **Figure SEC-3** the cutting plane line is drawn in the top view. Therefore, the front view location is now drawn as a full section. The cutting plane line is drawn with a phantom line and appears in the top view with the arrows at the end of the cutting plane line pointing in the viewing direction.

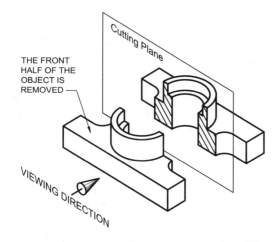

Figure SEC-1 –The cutting plane separates the object into two halves allowing the internal features of the rear half to appear.

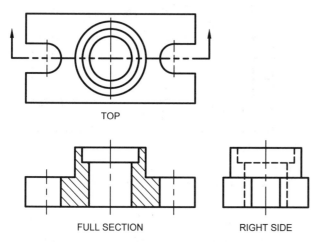

Figure SEC-2 –The front view appears as the Full Section.

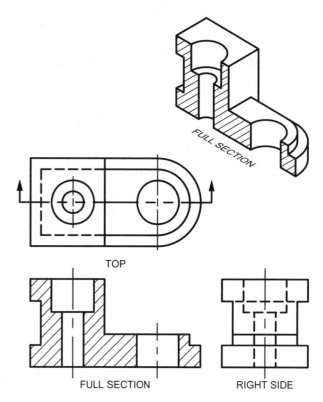

Figure SEC-3 – Full Section with isometric

Though half of the object is removed, this sectional view is called a ***full section*** since the cutting plane cuts all the way across the object. The internal features normally shown with hidden lines are now represented with visible lines. The portion of the object where the cutting plane slices through material is crosshatched with light section lines. There are many different types of lines representing various materials. Since our focus is to understand how the section view appears, we will use a traditional pattern of section lines at a 45° angle spaced approximately 1/8" apart. Where the cutting plane passes through a whole or slot (a void), section lines do not appear. Center lines for circular features are shown in the sectional view.

HALF SECTIONS

Half sections are useful for showing both the internal features and external features of an object. They are typically used with symmetrical objects and provide a quick visualization of both the inside and outside of the object. **Figure SEC-4** illustrates a half section.

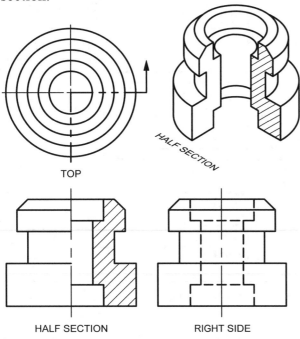

Figure SEC-4 – Half Section with isometric.

Although the cutting plane removes a fourth of the object, it is termed a ***half section*** because the cutting plane cuts only halfway across the object. In the sectional view, half of the object appears as a section and the other half appears as a normal view. Typically, hidden lines are not shown on the half section. Notice that the cutting plane has only one arrow. A center line is used to separate the half appearing in section (internal) from the half appearing as a typical orthographic view (external).

OFFSET SECTIONS

Not all objects contain holes and slots that are aligned in one plane. Quite often these voids are offset one from another. In order to pass the cutting plane through these holes and slots the cutting plane line must be bent line at 90° angles as in Figure **SEC-5**.

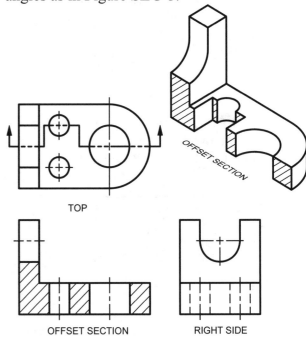

Figure SEC-5 – Offset Section with isometric

The cutting plane in **Figure SEC-5** could pass through either the front or rear hole in the mid-part of the object because the holes are in alignment and are the same diameter and depth. Notice that the bend in the cutting plane line does not create a visible line in the section view. If a line representing the bend in the cutting plane were included, the section would appear as separate pieces of material placed together. The cutting plane of the *offset section* has two arrows and both point in the same direction as the viewing direction.

REVOLVED AND REMOVED SECTIONS

Some objects have elongated features like handles, bars, spokes, and webs. Lines representing these features quite often appear parallel, or almost parallel, and become difficult to interpret in the orthographic views.

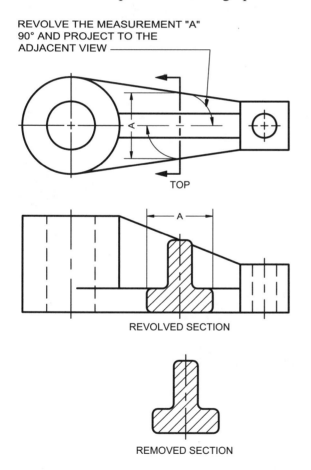

Figure SEC-6 – A Revolved Section is drawn on the front view. If this section is moved off the view, it becomes a Removed Section.

A *revolved section* shows a cutting plane perpendicular through the feature that will be drawn in the section. The material cut with the cutting plane line is then rotated 90° and the measurements are projected to the adjacent view. This revolved section can be drawn directly on the view as in **Figure SEC-6**. If there is not room or if the drawing becomes too cluttered, the section can be drawn adjacent to the view as a *removed section*. The removed section is similar to the revolved section because the cutting plane passes perpendicular to the elongated feature. The only difference between the revolved and removed sections is the location of the sectional drawing.

BROKEN - OUT SECTIONS

A *broken out section* avoids the use of a separate section drawing to clarify a particular internal feature. To illustrate the interior features, a portion of the object is imaginatively broken away and the interior features are drawn in section as in **Figure SEC-7**. A jagged line is used to indicate the break and separate the break from the normal exterior view. This broken out section can be as large or small as necessary to provide a clear view of the internal features.

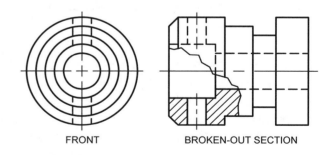

Figure SEC-7 – A Broken-Out Section.

ALIGNED SECTIONS

On some parts holes and special features, like ribs and webs, may not all be in alignment. The true orthographic projection of these features results in a distorted image of the object. However, if these features are revolved or rotated until they are in alignment with the cutting plane line and then projected, they more clearly represent the features of the object. Although an ***aligned section*** violates the principles of orthographic projection, alignment allows the features to be shown with true measurements from the central axis. The part shown in **Figure SEC**-8 is drawn as a full section. However, a half section could be used to illustrate the features of the object.

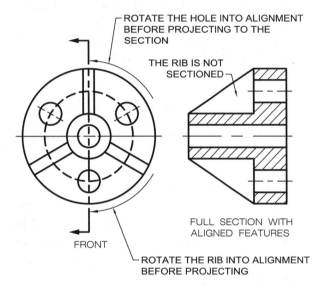

Figure SEC-8 – In the front view, the holes and rib are first revolved into alignment with the cutting plane and then projected to the sectional view.

A ***rib*** or ***web*** is a narrow, flat part that provides support to the object. Since these parts are thin, adding section lines creates the impression that they are as thick as the rest of the object. Omit sectioning the rib or web area in the sectional view when the cutting plane passes longitudinally through the web or the rib.

1.

Sketch the FRONT view as a FULL SECTION in each problem. The given views are complete.

TOP

FULL SECTION

RIGHT SIDE

A

2.

TOP

FULL SECTION

RIGHT SIDE

A

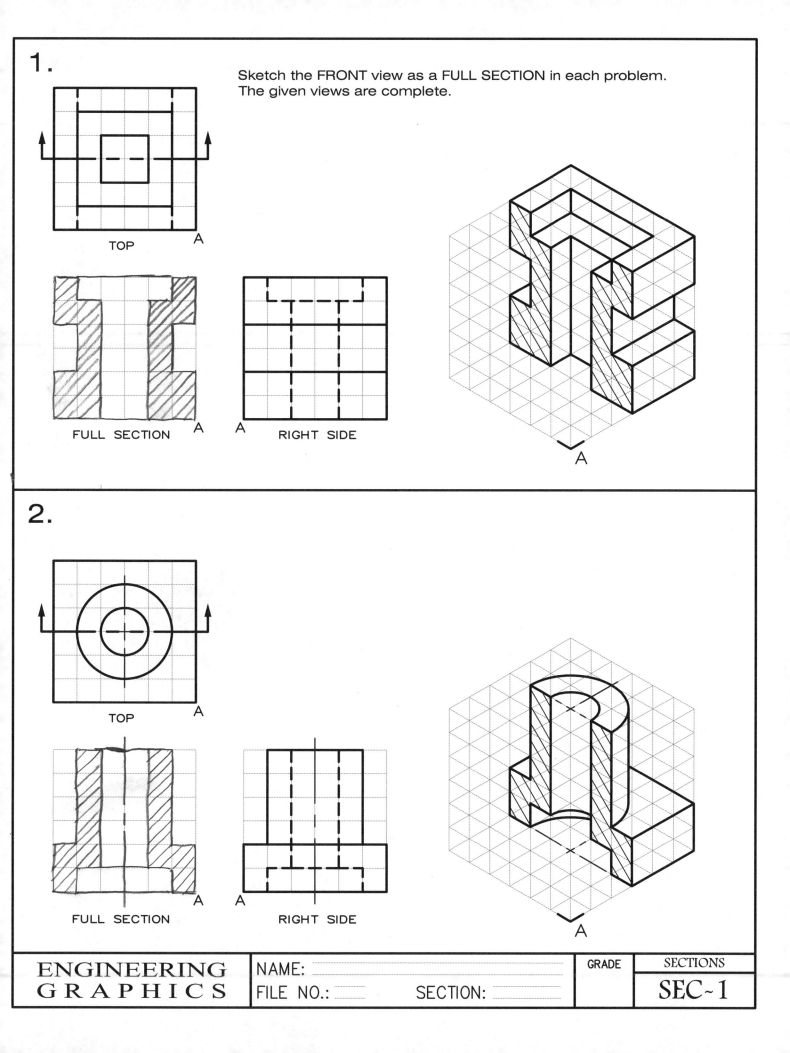

ENGINEERING GRAPHICS	NAME:	GRADE	SECTIONS
	FILE NO.: SECTION:		SEC~1

Sketch the front view as a HALF SECTION in each problem. Refer to the ISOMETRIC view of the section drawing to determine where the cutting plane slices through material.

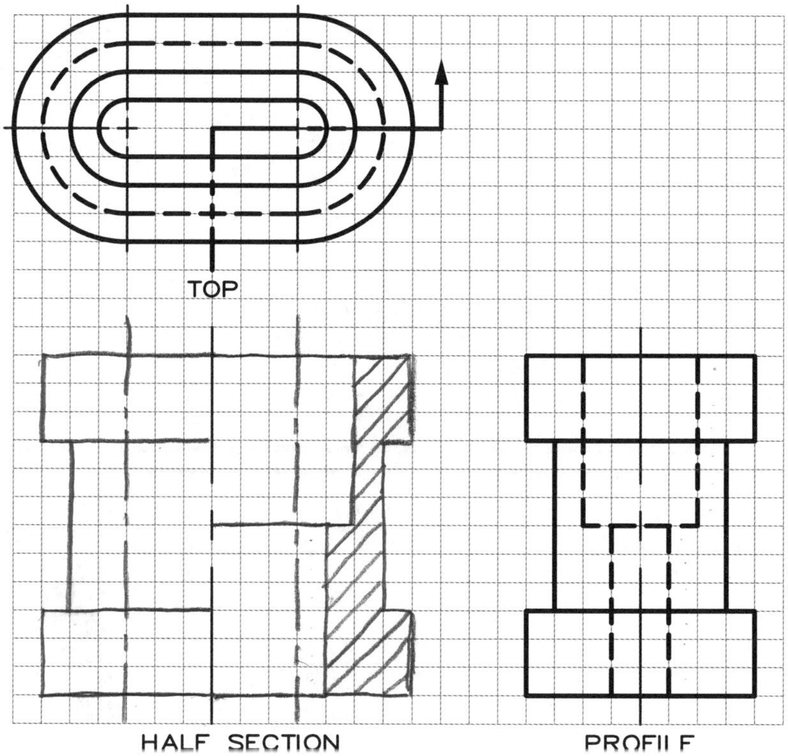

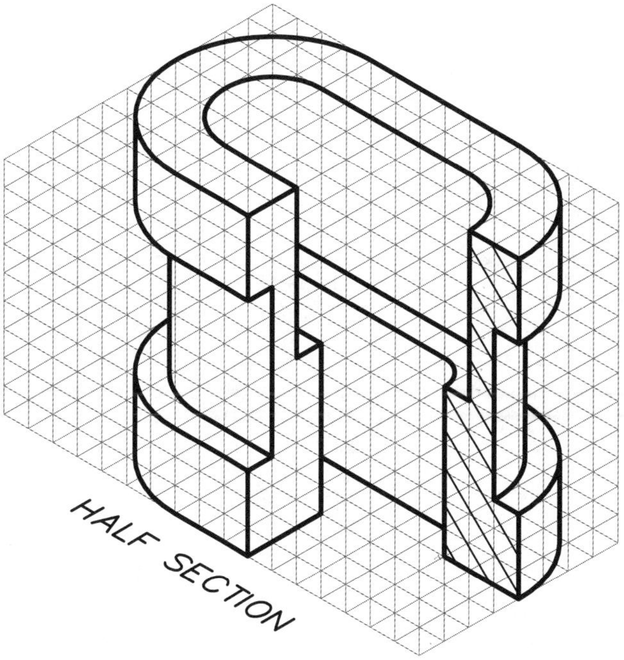

1.

TOP

HALF SECTION

PROFILE

HALF SECTION

2.

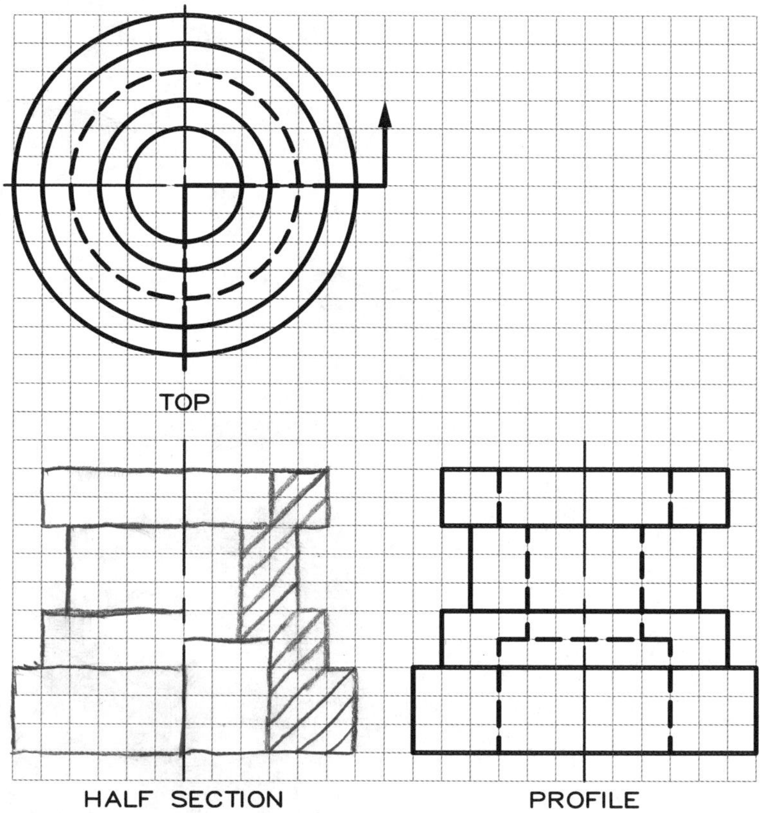

TOP

HALF SECTION

PROFILE

HALF SECTION

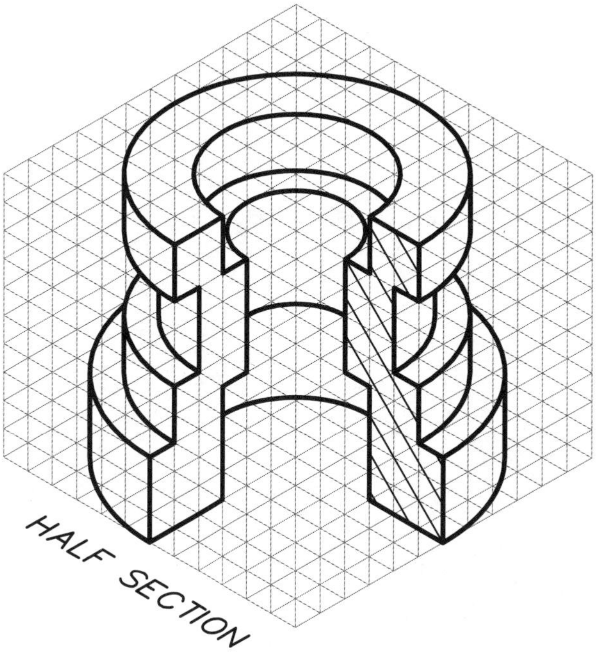

ENGINEERING GRAPHICS

NAME:

FILE NO.:   SECTION:

GRADE

SECTIONS

SEC-2

Sketch the front view as a FULL SECTION in each problem. Refer to the ISOMETRIC view of the section drawing to determine where the cutting plane slices through material.

1.

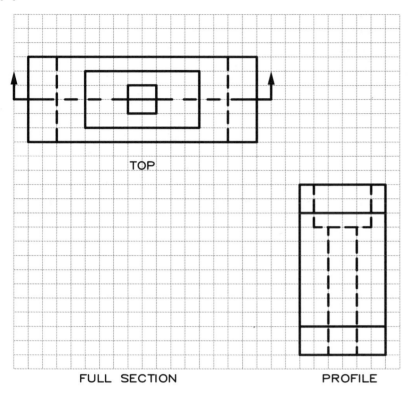

TOP

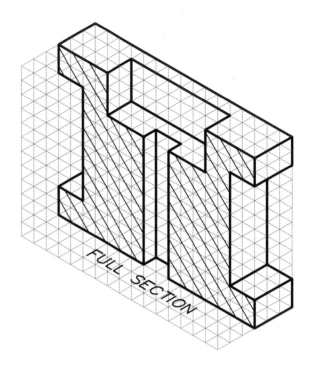

FULL SECTION

FULL SECTION   PROFILE

2.

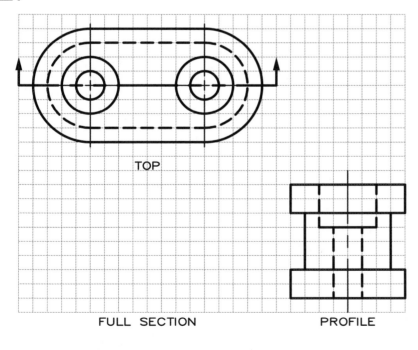

TOP

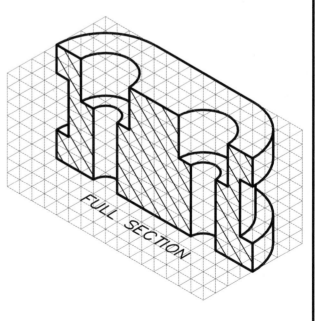

FULL SECTION

FULL SECTION   PROFILE

ENGINEERING GRAPHICS	NAME: _____ FILE NO.: _____ SECTION: _____	GRADE	SECTIONS
			SEC~3

Sketch the indicated SECTION view of each object. The
given ORTHOGRAPHIC views are complete.

1.

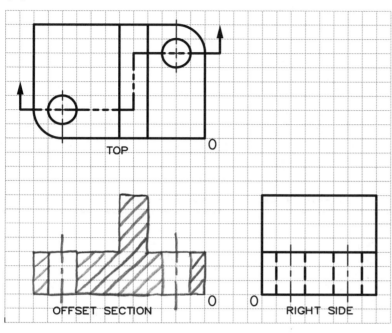

TOP

0

OFFSET SECTION

0

0

RIGHT SIDE

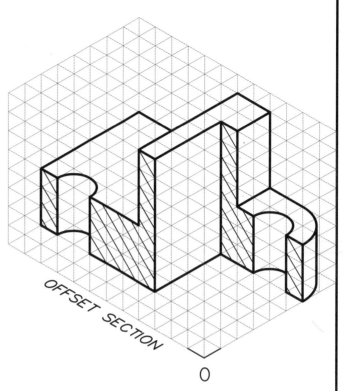

OFFSET SECTION

0

2.

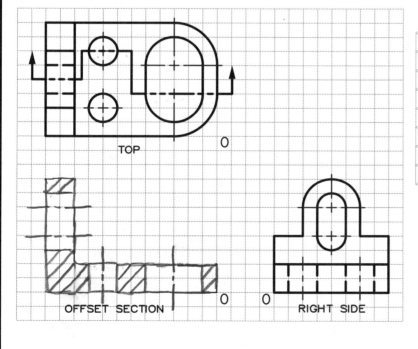

TOP

0

OFFSET SECTION

0

0

RIGHT SIDE

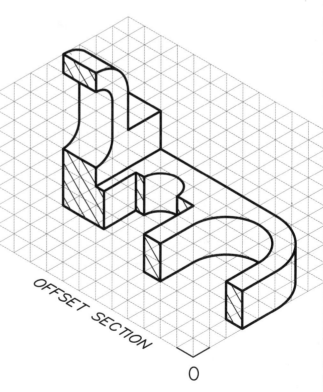

OFFSET SECTION

0

ENGINEERING
GRAPHICS

NAME: _____

FILE NO.: _____   SECTION: _____

GRADE

SECTIONS

SEC~4

Sketch the indicated SECTION view of each object. The given ORTHOGRAPHIC views are complete.

1.

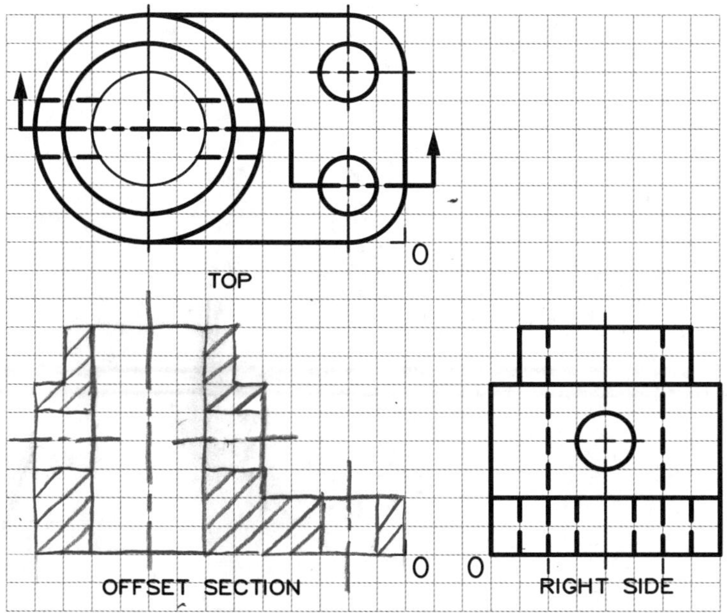

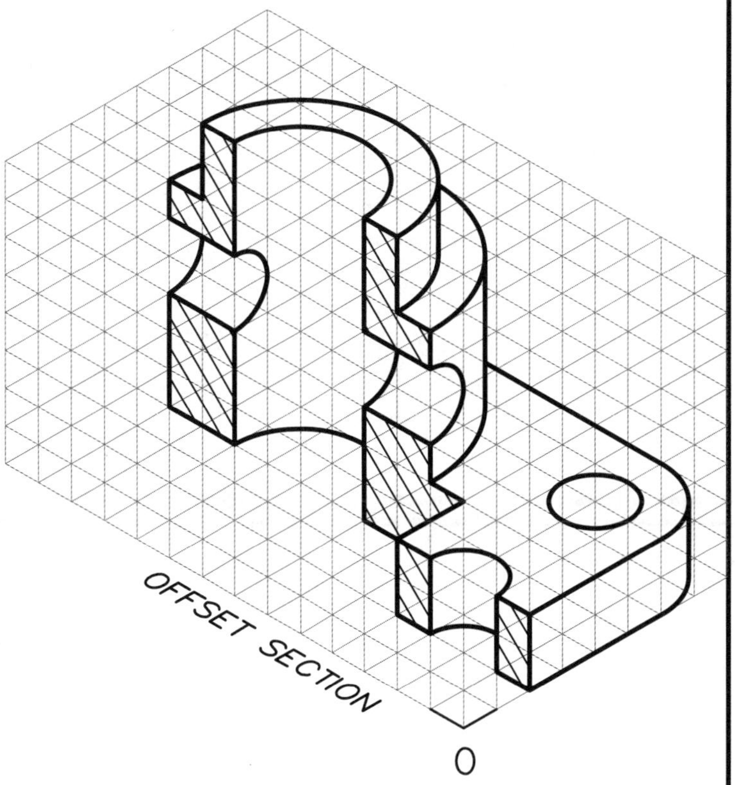

TOP

OFFSET SECTION

RIGHT SIDE

OFFSET SECTION

0

2.

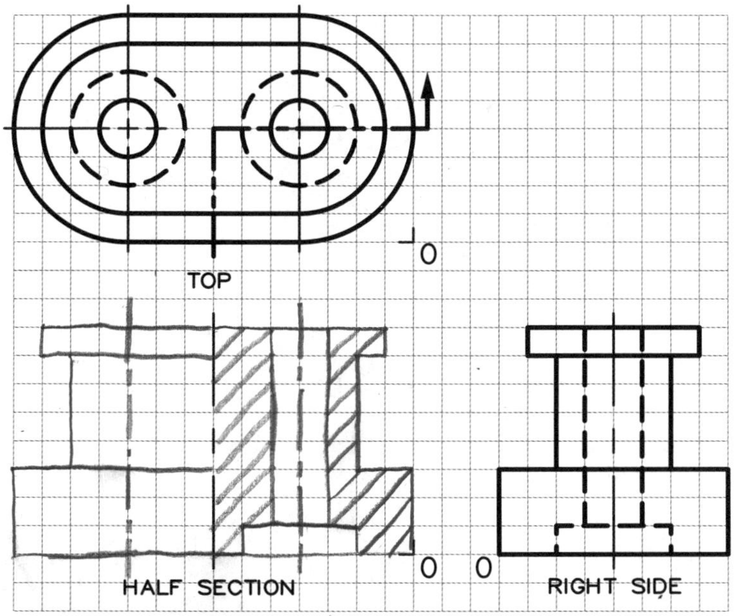

TOP

HALF SECTION

RIGHT SIDE

HALF SECTION

0

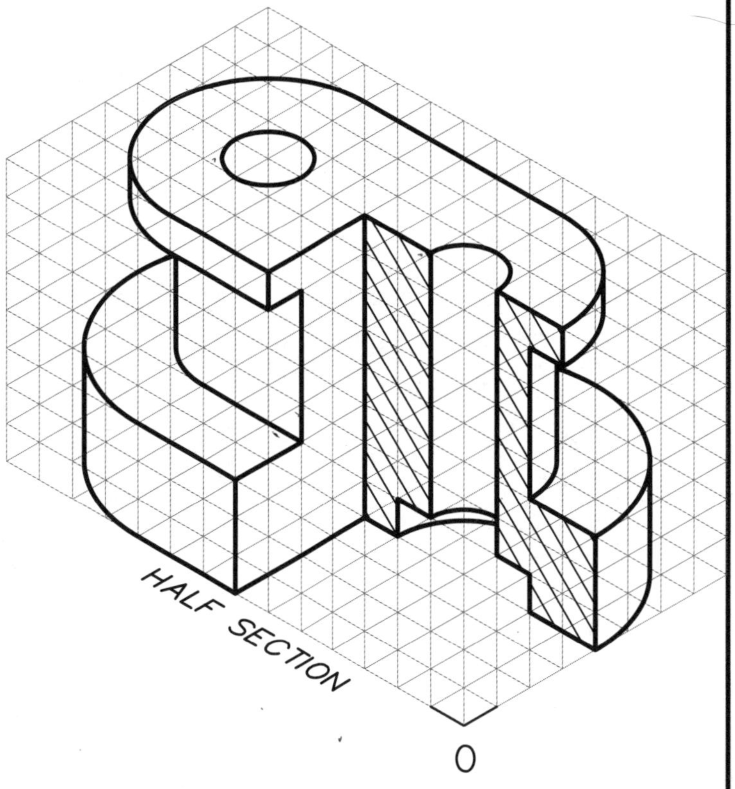

ENGINEERING GRAPHICS

NAME:

FILE NO.:   SECTION:

GRADE

SECTIONS

SEC~5

Sketch the FRONT view as a HALF SECTION for each problem.

1.

2.

3.

Note: Problem 3
has 2 possible
solutions.

4.

5.

6.

ENGINEERING
GRAPHICS

NAME:
FILE NO.:   SECTION:

GRADE

SECTIONS

SEC~6

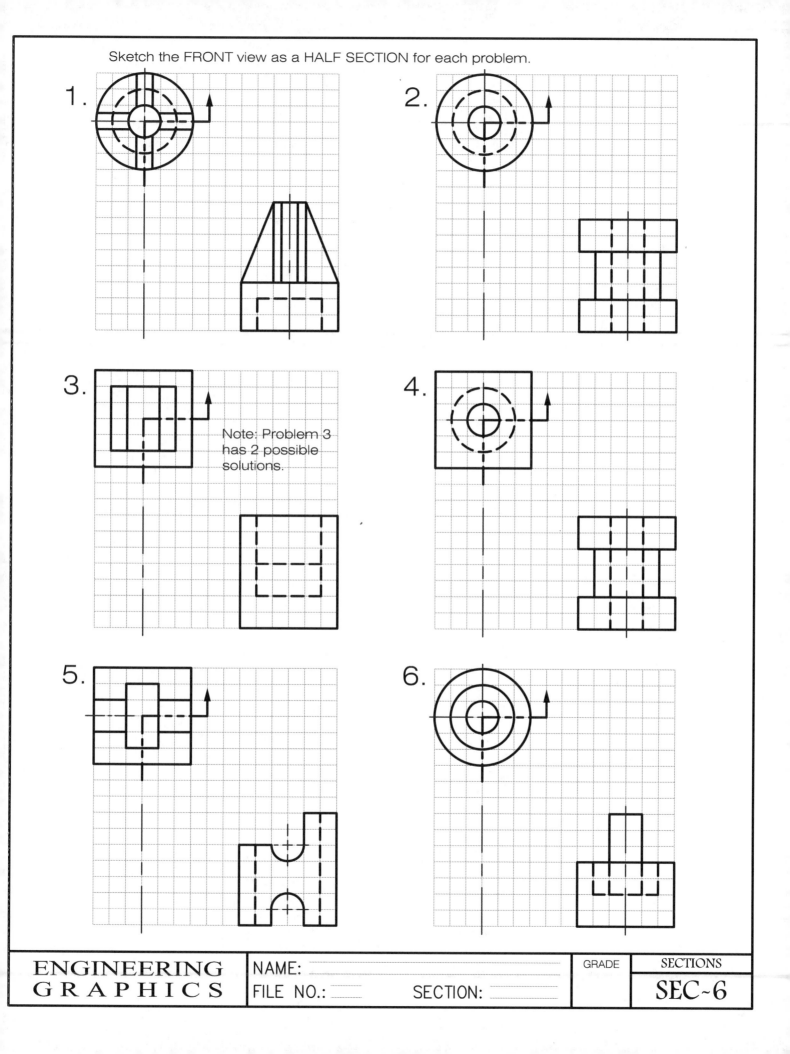

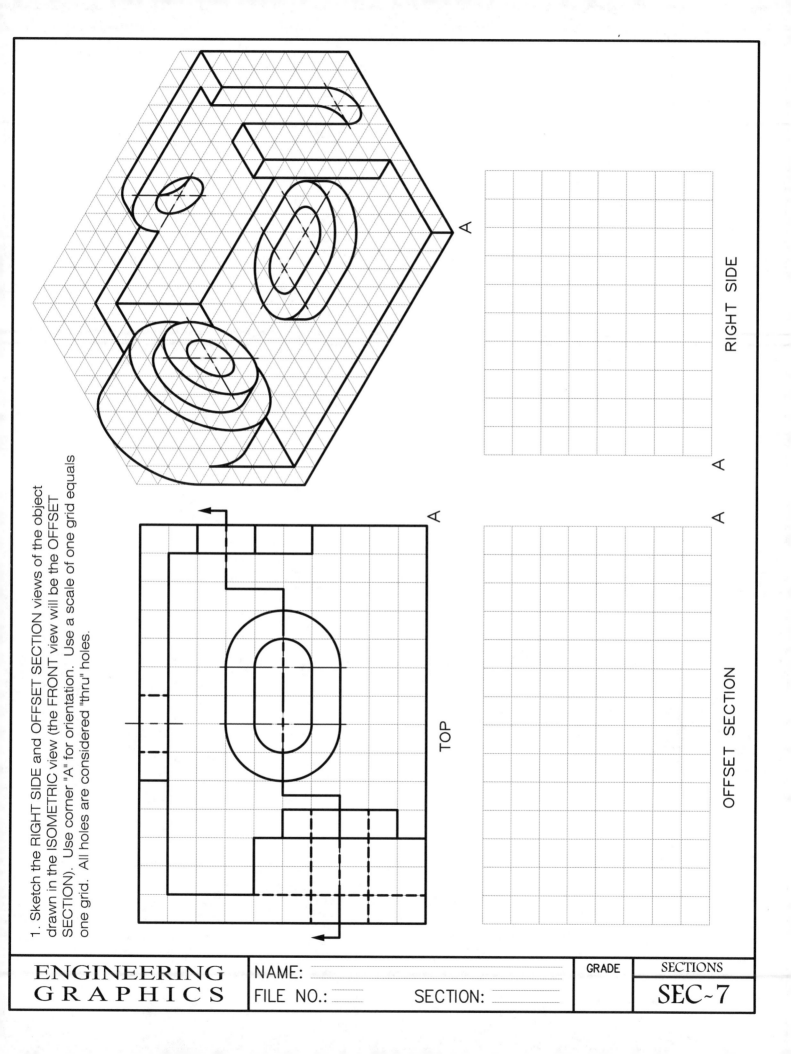

1. Sketch the RIGHT SIDE and OFFSET SECTION views of the object drawn in the ISOMETRIC view (the FRONT view will be the OFFSET SECTION). Use corner "A" for orientation. Use a scale of one grid equals one grid. All holes are considered "thru" holes.

RIGHT SIDE

A

A

A

TOP

A

A

OFFSET SECTION

ENGINEERING
GRAPHICS

NAME:

FILE NO.:   SECTION:

GRADE

SECTIONS

SEC~7

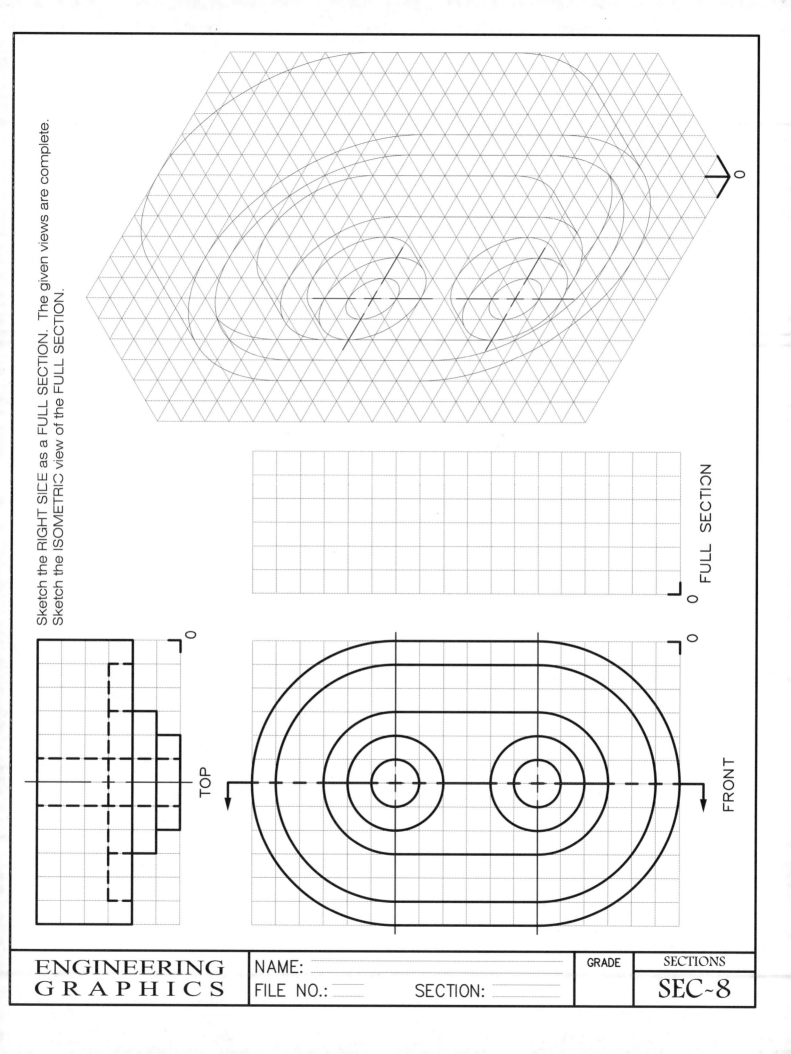

Sketch the RIGHT SIDE as a FULL SECTION. The given views are complete.
Sketch the ISOMETRIC view of the FULL SECTION.

TOP

FULL SECTION

FRONT

ENGINEERING
GRAPHICS

NAME:

FILE NO.:   SECTION:

GRADE

SECTIONS

SEC~8

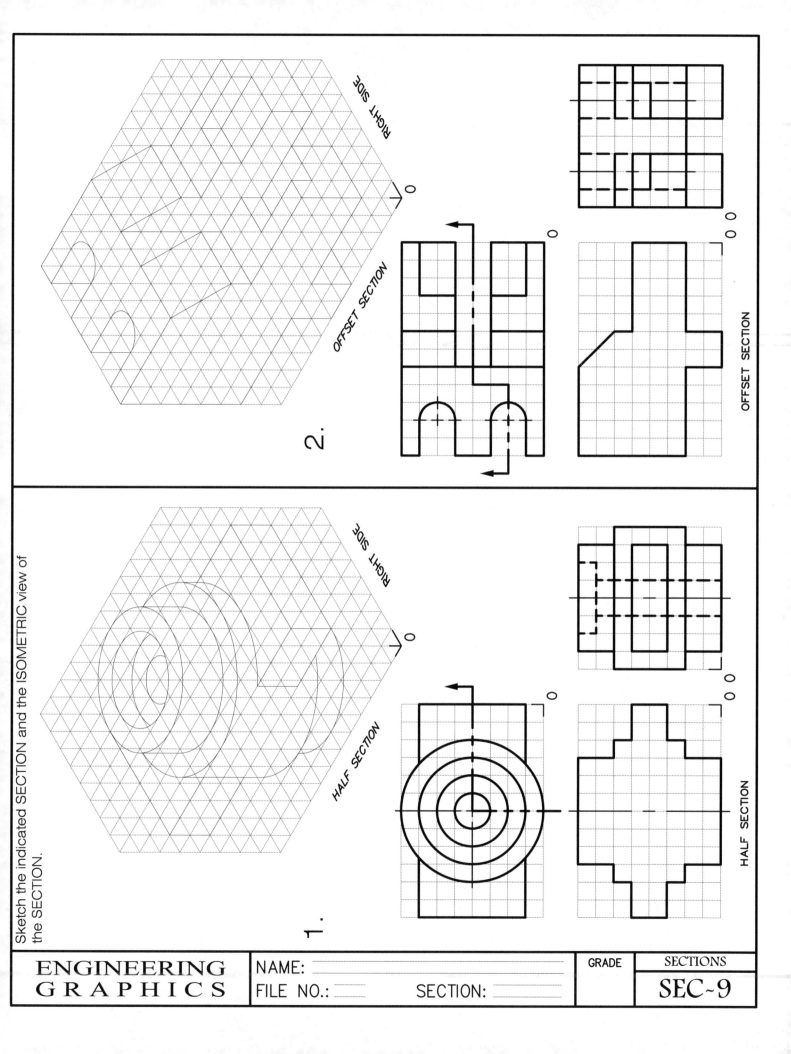

Sketch the indicated SECTION and the ISOMETRIC view of the SECTION.

RIGHT SIDE

OFFSET SECTION

2.

OFFSET SECTION

OFFSET SECTION

RIGHT SIDE

HALF SECTION

HALF SECTION

HALF SECTION

1.

ENGINEERING
GRAPHICS

NAME:

FILE NO.:   SECTION:

GRADE

SECTIONS

SEC~9

Sketch the HORIZONTAL VIEW as an
OFFSET SECTION. Draw the cutting
plane line indicating where the object
is cut.
The given views are complete.
Sketch the ISOMETRIC view of the
OFFSET SECTION. Note the
orientation of point "O".

O

OFFSET SECTION

LEFT SIDE

O

O

FRONT VIEW

O

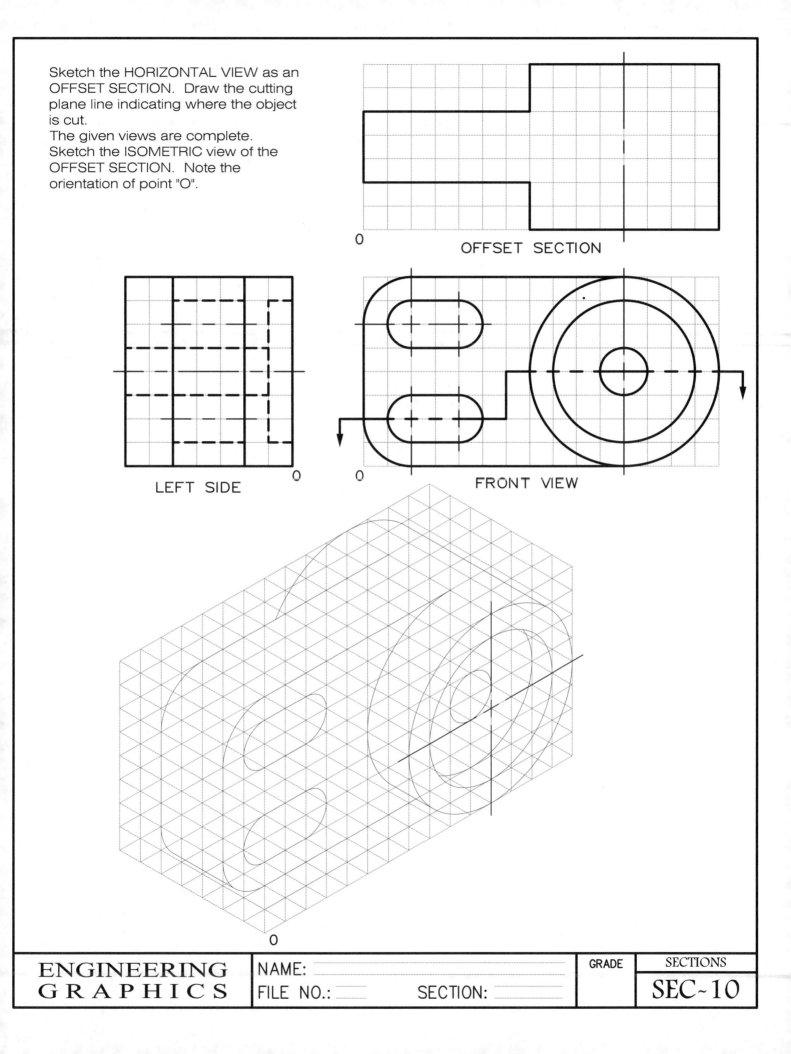

ENGINEERING
GRAPHICS

NAME:

FILE NO.:   SECTION:

GRADE

SECTIONS

SEC~10

Sketch the indicated SECTIONS.

1.

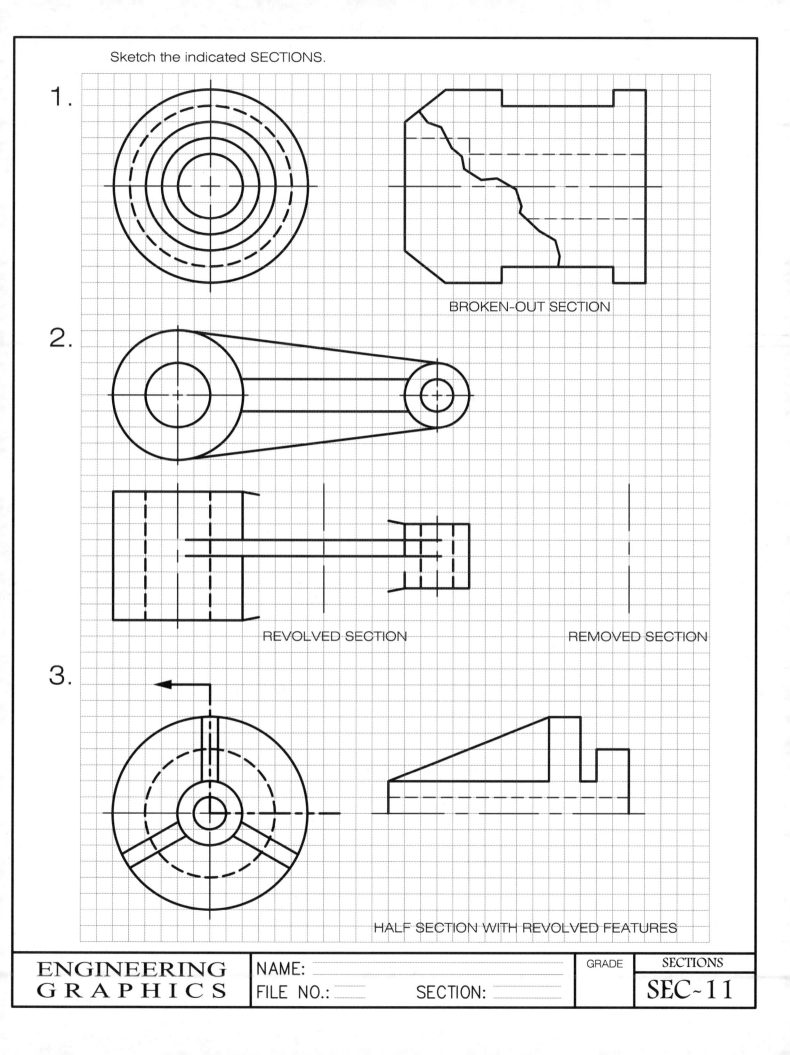

BROKEN-OUT SECTION

2.

REVOLVED SECTION

REMOVED SECTION

3.

HALF SECTION WITH REVOLVED FEATURES

| ENGINEERING GRAPHICS | NAME: | GRADE | SECTIONS |
| | FILE NO.: SECTION: | | SEC~11 |

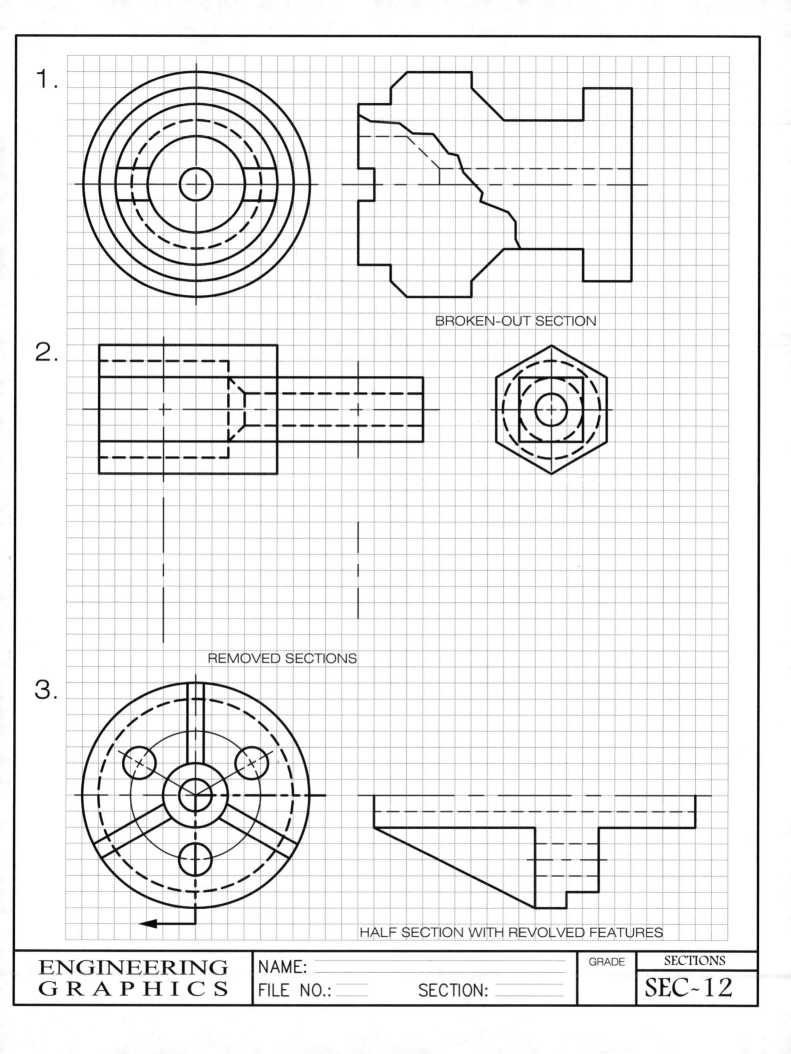

1.

BROKEN-OUT SECTION

2.

REMOVED SECTIONS

3.

HALF SECTION WITH REVOLVED FEATURES

ENGINEERING
GRAPHICS

NAME: _____

FILE NO.: _____ SECTION: _____

GRADE

SECTIONS

SEC~12

Sketch the indicated SECTIONS. Sketch all missing lines.

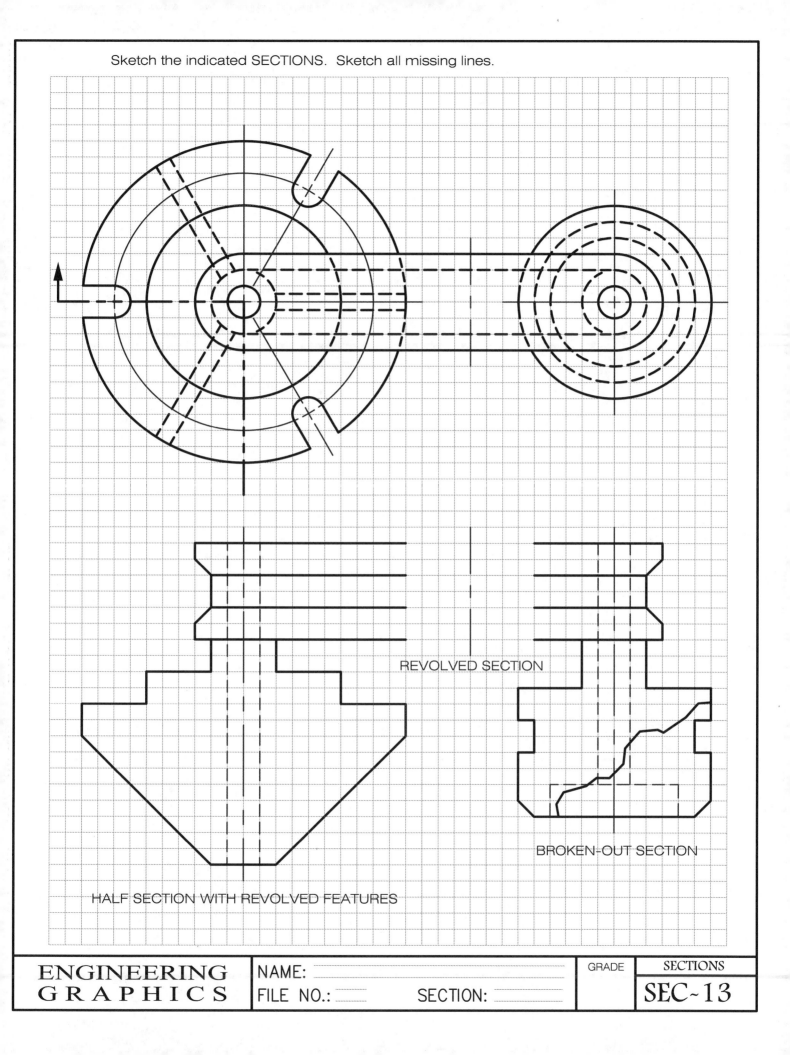

REVOLVED SECTION

BROKEN-OUT SECTION

HALF SECTION WITH REVOLVED FEATURES

ENGINEERING
GRAPHICS

NAME:
FILE NO.:   SECTION:

GRADE

SECTIONS

SEC~13

DIMENSIONING

Dimensions are important because they describe the exact size of the various parts of the object. To draw a dimension, start with a pair of extension lines. These extension lines are aligned with each edge of the object or a part of the object. While drawing the extension lines, allow approximately a 1/8" gap between the object and the beginning of the extension line. Draw the extension line approximately 3/8" long.

Next, determine the location of the dimension line. **Figure DIM-1** shows that the dimension line should be approximately 3/8" from the object and the extension lines should extend 1/8" beyond the dimension line. The actual measurement is placed on the dimension line with a break in the line to allow for placement of the text. The text is also 1/8" high. The arrows on each end of the dimension line are very slender and approximately 1/8" long.

Leaders are used for some notes to indicate special features, diameters or radii. The arrowhead should always touch the edge of what is being noted, and, in the case of the diameter, it should point toward the center of the circle while touching the edge of the circle representing the hole.

Typically, machine drawings use one of two types of dimensioning: either the English system with its base unit being the inch or the Metric system (Systems International) with the basic unit being the millimeter. In order to allow the use of either system, this workbook is based on a grid system where 1/8" approximates 3 mm. This system makes it easy to count grids, multiply and place the measurement.

Specific rules apply for the notation of each system. For fractional inches there is no leading zero before the decimal point. As an example, a quarter of an inch becomes .25 and not 0.25. Avoid inserting a closed quotation mark for an inch measurement such as .25", and using the notation of "mm" for millimeter measurements such as 68mm.

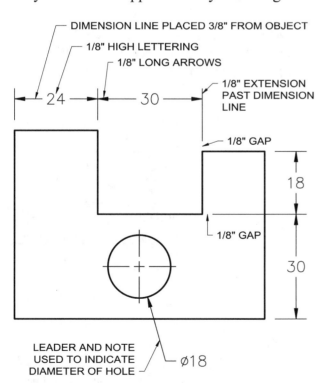

Figure DIM-1 – Placement of linear dimensions.

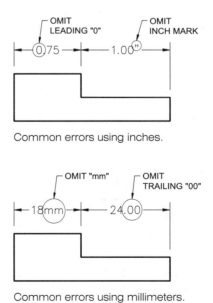

Figure DIM-2 – Notation errors.

In the United States, placement of dimensions is guided by a set of rules that are the standard set by the American National Standards Institute (ANSI). Most companies also have common practice standards that are used by a particular industry. Machine drawings are dimensioned so that all the text appears in one direction (uni-directional) and can be read when the sheet is oriented so that the bottom of the page is at the bottom of the table.

Basic rules are followed to properly place dimensions for greatest clarity and to avoid superfluous dimensions. These rules are discussed in order and compiled on Plate DIM-15 for quick reference.

BASIC RULES FOR DIMENSIONING

The goal of this workbook is to introduce a student to the basics of dimensioning and will concentrate on *continuous dimensioning* that starts on one edge of an object and continues across the object. Unless the object has a finished surface, the dimensions can start on either end or the top or bottom. Omit one dimension in a series since the overall dimension will provide the missing measurement.

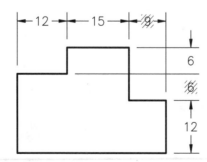

Figure DIM-3

Omit one dimension in a linear series.
Keep the dimensions "continuous" or "in line".

Always place dimensions where the shape is most clearly visible. The cuts in the object in **Figure DIM-4** are more easily distinguished in the front view. The corners of the cuts in

the front view also appear to be formed with "L's" where lines intersect. The same cuts, when projected to the top view, appear to form corner intersections with "T's". In order to determine the most visible view, always dimension to the "L's" and not the "T's".

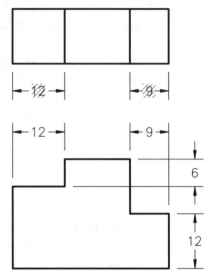

Figure DIM-4

Omit one dimensions in the top view since the features are more clearly visible in the front view.

Never dimension to a hidden line unless a peculiarity of the object forces the location of an extension line on a hidden line.

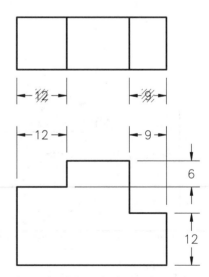

Figure DIM-5

Omit both dimensions in the top view since the features are more clearly visible in the front view.

If possible, always keep dimensions off the object. Sometimes, as in the case of radii, it is more legible to place a dimension on the object.

Avoid superfluous dimensions. For example, after a dimension for overall width is placed, do not repeat the width dimension in another view.

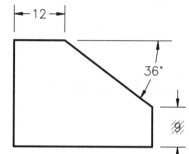

Figure DIM-6

For an angled surface, use either two coordinates or a coordinate and angle.

SIZE

In order to logically place dimensions, approach the object with the intent of dimensioning in the order of *size, location* and *overall dimensions*. Start with the smaller parts of the object and work toward the larger and finally the overall dimensions. Sizing is the process of providing the measurements for the individual parts of the object. Remember to place size dimensions where the part is most visible.

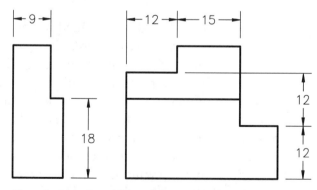

Size the parts of the object where they are most clearly visible.

Figure DIM-7

LINEAR DIMENSION PLACEMENT

Often, dimensioning the smallest features becomes a challenge to arrange the arrows and measurements for clarity. Where room is sufficient, the first item to be included is the text. For a slightly larger feature, arrows are included with the text. Notice that some arrows are assumed where small measurements are grouped together.

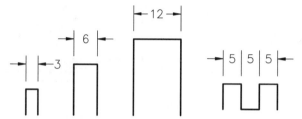

For very small dimensions, the arrows and text are placed outside the extension lines. As room permits, the text is placed betwen the extension lines and finally the text and arrows. The last illustration shows a series of small measurements where the "interior" arrows are omitted.

Figure DIM-8

ARC DIMENSION PLACEMENT

Small arcs are similar to linear dimensions since the smallest arcs omit the center mark and place the arrow and text on the outside of the curve. As the radius increases, the arrow moves inside the curve and the radius is noted with a leader. When the radius is large enough, the measurement also moves inside the curve.

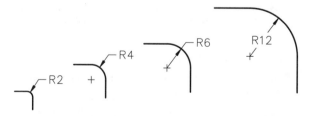

As the radius becomes progressively larger, include the radius point, followed by the line and arrow and finally the text.

Figure DIM-9

HOLES AND CYLINDRICAL FEATURES

Holes and cylindrical features are represented by a circle in one view and either visible lines or hidden lines in the adjacent views. Although similar, they are dimensioned differently. The hole is dimensioned where it appears round or *circular*. The arrow attached to the leader touches the outside of the circle and points to the center of the hole.

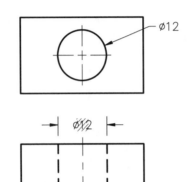

Figure DIM-10

Size holes where they appear circular. Never dimension to the hidden lines.

The cylinder is dimensioned where it appears *rectangular*. Since the hole is a complete circle, always provide the diameter and never the radius. Note that the measurement is a diameter by using the symbol of the circle with a slash.

Figure DIM-11

Size cylinders where they appear rectangular and not where they appear circular.

MACHINED HOLES

Sizes of machined holes, formed by first drilling a hole and then counter boring, spot facing or counter sinking, are given in the form of a note. Since these are holes, the note is placed where the counter bore, spot face and countersink appear circular. These holes require special notation with symbols being

used to represent the machining operation. Note that the depth is also included as a part of the note. A *drilled hole* that goes all the way through the material requires only a diameter indicated by symbol of the circle with the slash. If the hole does not go all the way through the material, it is termed a *blind hole* and includes the depth symbol (indicted by the bar and downward pointing arrow).

The *countersink* note indicates the diameter of the drilled hole and the symbol "V" to indicate the countersink. The larger diameter of the counter sink is noted and followed by the angle of the "V", typically 90° or 82°.

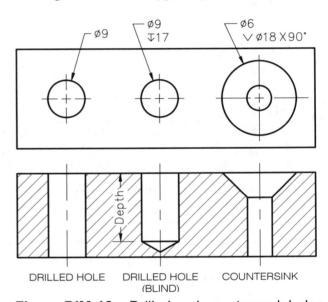

Figure DIM-12 – Drilled and countersunk holes.

The counterbored and spot faced holes are similar. The *counterbored hole* is made in two operations. First, the hole is drilled. Then the counterbore tool, which has a flat bottom, removes material to the depth indicated for the counterbore. This larger void, above the drilled hole, provides space for the head of a bolt to be recessed below the top surface of the material. A *spot face* is a similar machining operation. However, the spot face simply provides a smooth surface on a cast part for a secure gripping surface for a bolt head.

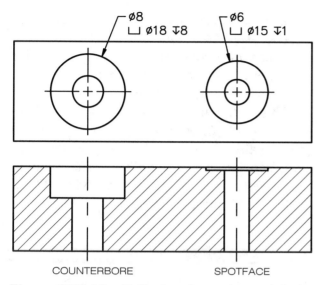

COUNTERBORE   SPOTFACE

Figure DIM-13 – Drilled and countersunk holes.

LOCATE

Location dimensions are used to locate features on the object. They are always placed where the feature can be dimensioned in two directions such as width and height. Location dimensions for circular features such as holes and cylinders are always placed on the object where the hole or cylinder appears round (circular) and not in the view where the hole appears as hidden lines.

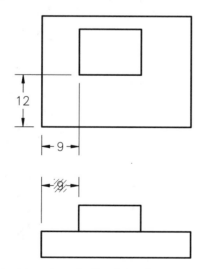

Locate the rectangular prism in the top view where two dimensions (height and width in this case) can be used and the position is more clearly visible.

Figure DIM-14 – Locating rectangular features.

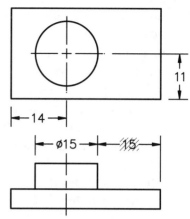

Locate the cylinder where it appears circular. The diameter is provided where the cylinder appears rectangular. Do not dimension to the edge of the cylinder where it appears rectangular.

Figure DIM-15 – Locating circular features.

FINISHED SURFACES

Finished surfaces result from the machining of a cast surface to provide a smooth surface for a specific purpose. This machined surface may be required to mate with another part or may be machined to fit within a certain tolerance. The finished surface is important while dimensioning as the linear dimensions typically begin from the finished surface. The finished surface is noted by the finished surface mark ("V") appearing where the finished surface appears as an edge view.

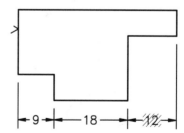

Always dimension from an edge designated as a finished surface. The finished surface mark will be located where the finished surface appears as an edge.

Figure DIM-16 – Dimensioning from a finished surface.

SLOTTED HOLES

Slotted holes can be dimensioned several different ways.

a. Center points on the longitudinal axis are located and the radii are indicated with a note.

b. Linear measurements indicate the overall size of the slot and the radii are specified.

c. A note indicates the two linear dimensions of the slot and another note specifies the radii.

Choose the most appropriate technique for sizing the slotted hole.

LOCATING HOLES OR CYLINDERS ON AN ARC OR CIRCLE

Holes or cylinders positioned in a circular pattern can be located by referencing the diameter of the circle or the radius of the arc. This technique is known as a *circle of centers* or *bolt circle* because the holes would typically allow for a bolt to pass through and fasten this part to another. The diameter of the bolt circle is drawn in the view where the bolt circle appears circular.

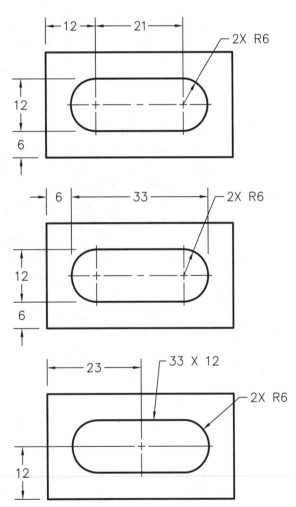

Each technique for dimensioning slotted holes has unique features. Avoid mixing the locational and sizing features of various techniques.

Figure DIM-17 – Techniques for sizing and locating slotted holes.

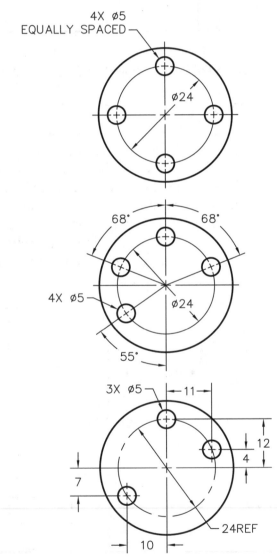

The bolt circle is used to locate holes about a common center.

Figure DIM-18 – Examples of locating circular features using a bolt circle.

After the diameter of the bolt circle is given, the location of the holes can be indicated by several methods.

a. If the holes are all equally spaced, the abbreviation "EQ SP" is added to the note giving the diameter and quantity of the holes.

b. The angles between the center lines of the holes can be provided with reference to only one of the centerlines.

c. List the bolt circle as a reference diameter by including the measurement in parenthesis or adding the notation "REF" after the measurement. Next, add the horizontal and vertical measurements from the center lines to the center mark of each hole.

OVERALL DIMENSIONS

After the size and location dimensions have been positioned on the object and necessary notes have been added, the overall dimensions are placed. The *overall dimensions* provide the overall height, width, and depth for the object. They are typically placed between the views of the object. Avoid indicating each of the overall dimensions more than once.

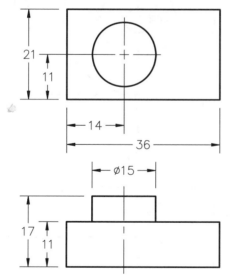

Overall dimensions are added for total height, width and depth. Place these between views where possible.

Figure DIM-19 – Overall dimensions.

It is permissible for extension lines to cross. However, never cross a dimension line with another dimension line or extension line as it breaks the dimension line and provides an opportunity for misinterpretation of the measurement.

ROUNDED END OBJECTS

In the case of rounded end objects, eliminate the overall dimension in the view where the object's rounded end appears circular. If an object has a rounded end, dimension to the center of the radius or the center of the cylindrical feature. Avoid giving an overall measurement in that view since a combination of the location dimension and the radius (rounded end) or diameter (cylindrical feature) provides the overall dimension.

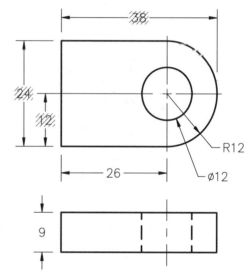

Omit overall dimensions for rounded end objects in the view where the rounded end appears circular.

Figure DIM-20 – Object with one rounded end.

Some objects are designed with two rounded ends and require different techniques depending on the precision required during manufacture. The first technique is similar to the technique for an object with one rounded end. The location dimension between the center points is given in the view where the ends appear circular and the measurements of the radii are given in a note. An overall width dimension is omitted since the location dimension and the radii provide the overall measurement.

The second technique is to provide overall dimensions in the view where the rounded ends appear circular. These dimensions provide the overall measurement from end to end and also the measurement from side to side. Since the overall size has been given, the radii are noted with only the "R" to indicate that the ends are formed with a constant radius. The correct technique depends on the application and the type of industry.

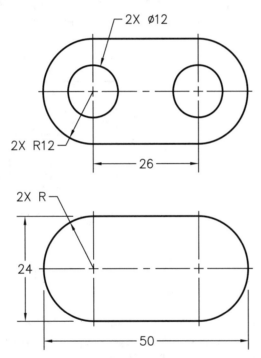

Objects with two rounded ends are dimensioned with a location dimension or an overall dimension.

Figure DIM-21 – Dimensioning rounded end objects.

The rules of dimensioning require a careful analysis of the object. Begin by making a rough sketch of probable placement of the dimensions. As dimensions are added, ways to simplify the number and presentation of the

dimensions will become apparent. **Figure DIM-22** provides a good illustration of how few dimensions are necessary to properly dimension an object. The cylindrical surface in the top view is located from the right edge. Because it appears rectangular in the front view, the diameter of the cylinder is given in the front view. The size of the counter sunk hole is given in the top view with a note. A radius is provided to the notch on the right side. This radius is self locating since the center point is on an edge and aligns with the centerline of the cylinder. The height in the front view must start from the bottom surface since this a finished surface. The overall width measurement is omitted due to the rounded end. The overall depth measurement is unnecessary as the diameter of the cylinder is given in the front view. The overall height is provided in the front view.

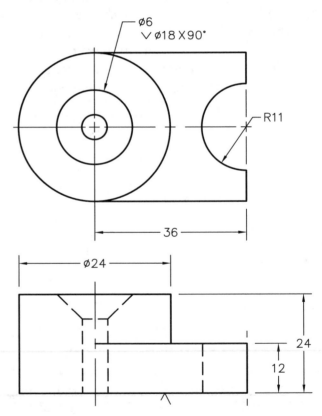

An object with a variety of features requiring careful consideration of dimension placement.

Figure DIM-22 – An object with several circular features.

DIMENSION EACH OBJECT. USE CONTINUOUS DIMENSIONING WITH SPECIAL ATTENTION
TO WHERE FEATURES APPEAR MOST VISIBLE. INCLUDE OVERALL DIMENSIONS FOR
WIDTH, HEIGHT AND DEPTH. OMIT ACTUAL MEASUREMENTS.

1.

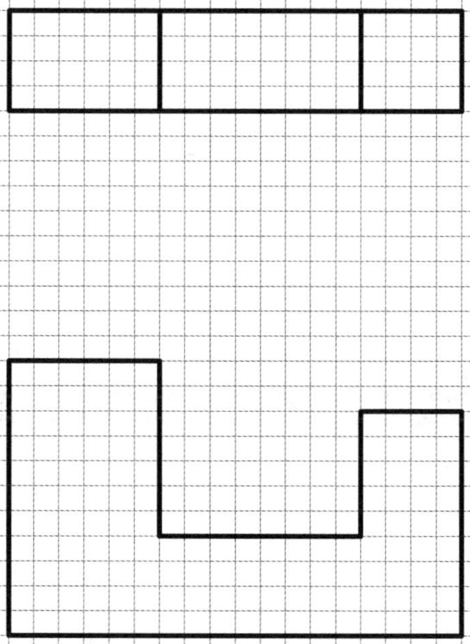

2.

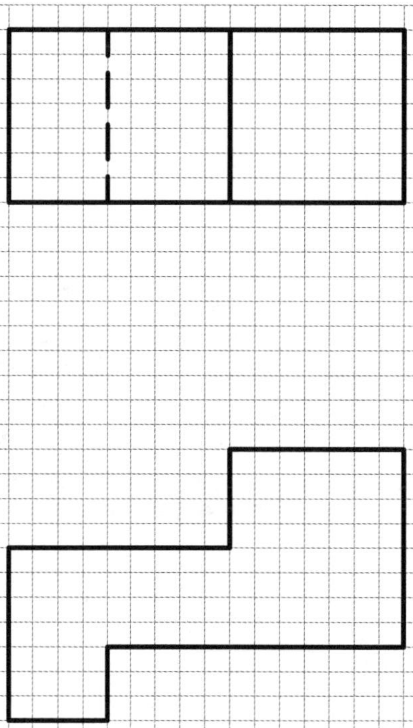

3.

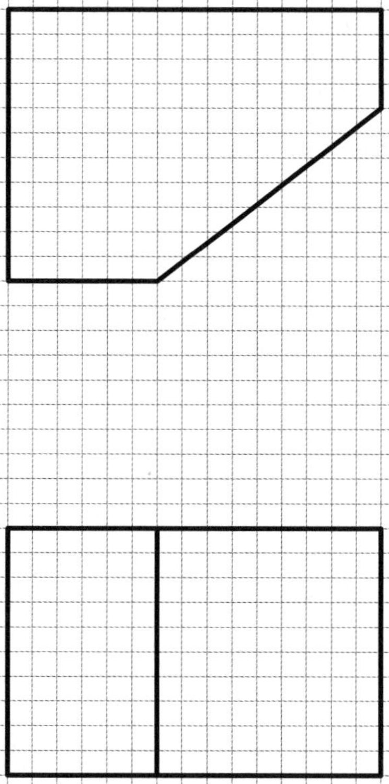

4.

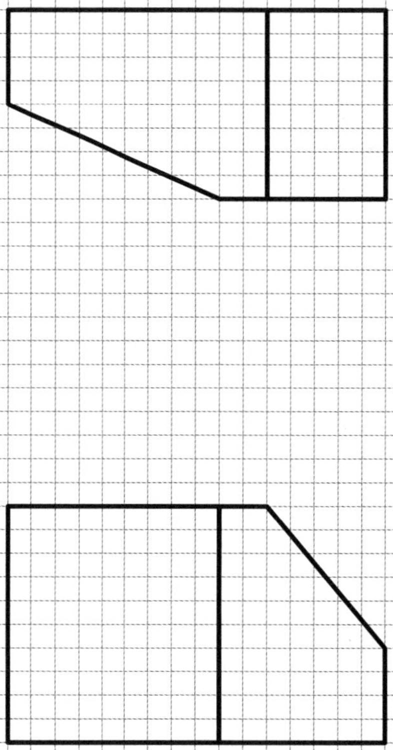

ENGINEERING
GRAPHICS

NAME:

FILE NO.:   SECTION:

GRADE

DIMENSIONING

DIM~1

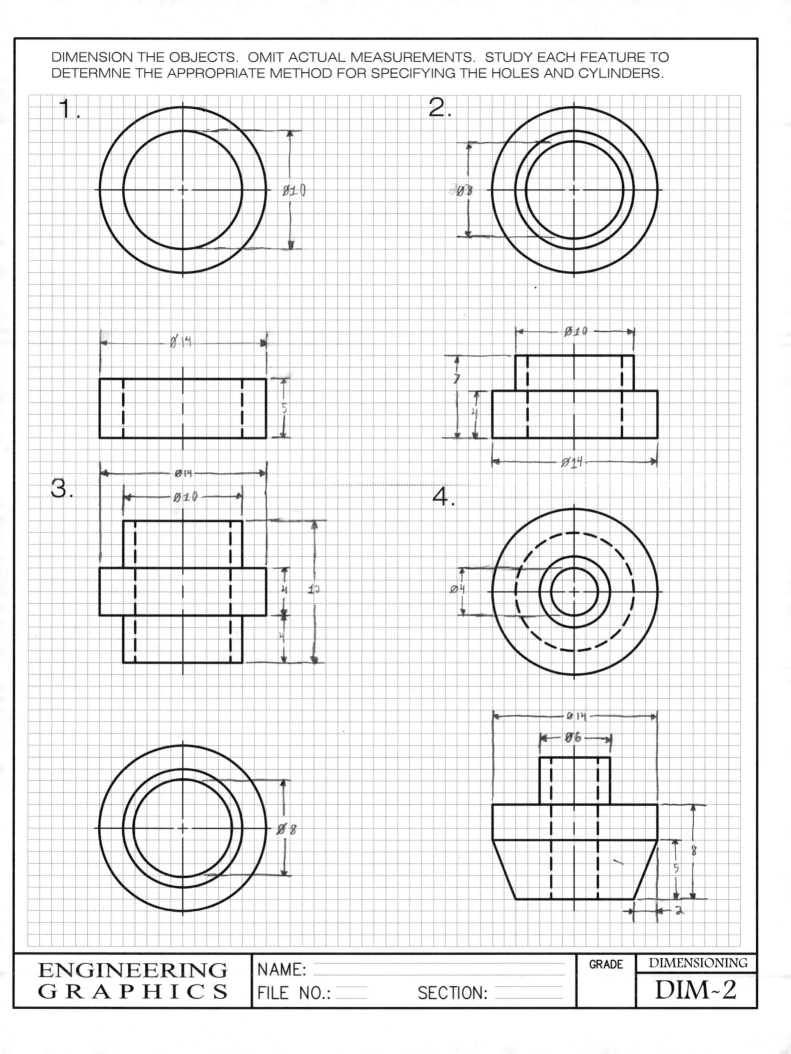

DIMENSION THE OBJECTS. OMIT ACTUAL MEASUREMENTS. STUDY EACH FEATURE TO DETERMNE THE APPROPRIATE METHOD FOR SPECIFYING THE HOLES AND CYLINDERS.

1.

Ø10

Ø14

5

2.

Ø8

Ø10

7

4

Ø14

3.

Ø14

Ø10

4

12

4

Ø8

4.

Ø4

Ø14

Ø6

8

5

2

ENGINEERING
GRAPHICS

NAME:

FILE NO.:   SECTION:

GRADE

DIMENSIONING

DIM~2

DIMENSION EACH OBJECT. USE CONTINUOUS DIMENSIONING WITH SPECIAL ATTENTION
TO WHERE FEATURES APPEAR MOST VISIBLE. INCLUDE OVERALL DIMENSIONS FOR
WIDTH, HEIGHT AND DEPTH. INCLUDE ACTUAL MEASUREMENTS.

1.

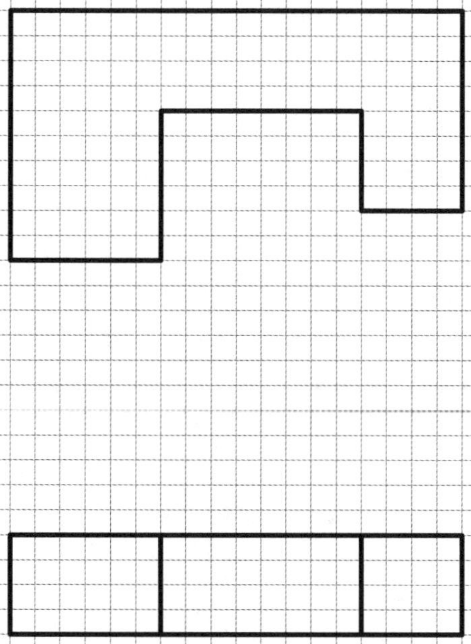

2.

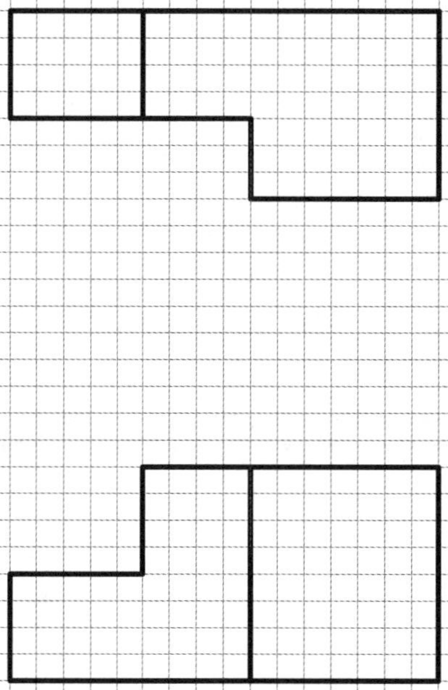

3.

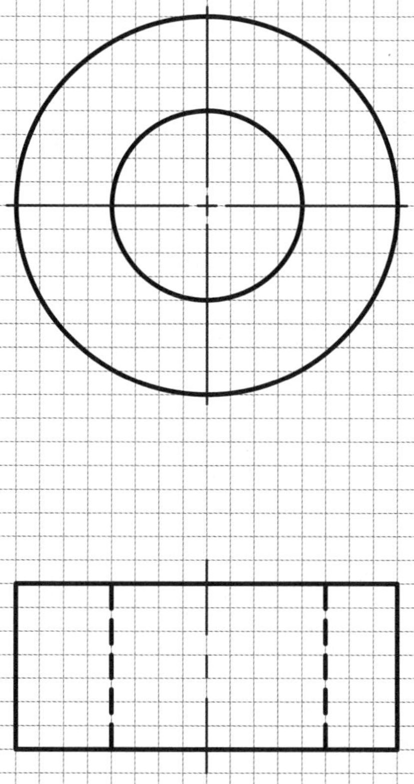

4.

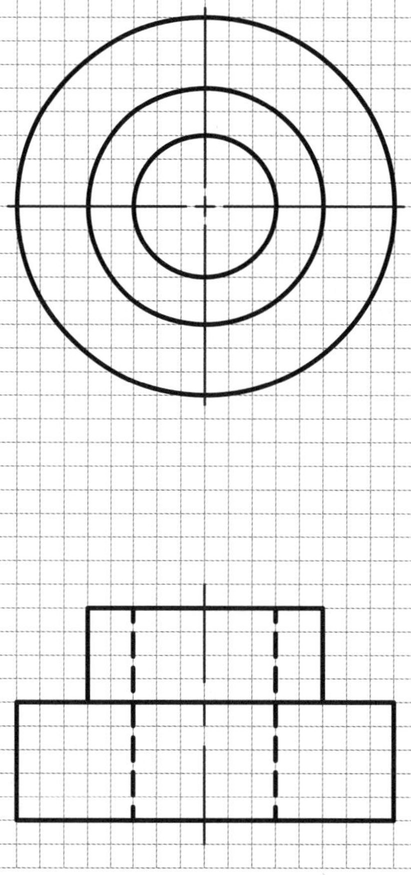

ENGINEERING
GRAPHICS

NAME:

FILE NO.:   SECTION:

GRADE   DIMENSIONING

DIM~3

DIMENSION THE OBJECTS. EACH GRID IS EQUAL TO 3mm OR 1/8". USE THE APPROPIATE
TECHNIQUES TO SIZE AND LOCATE THE FEATURES AND PROVIDE OVERALL DIMENSIONS.

1.

2.

3.

4.

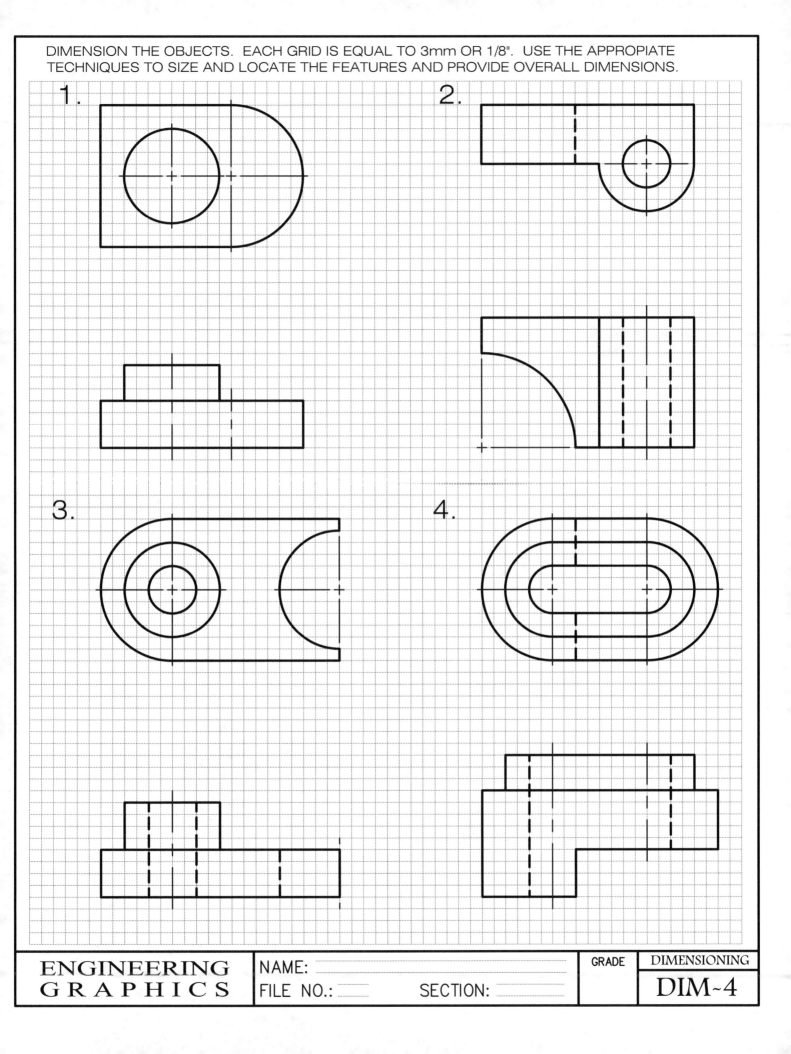

ENGINEERING
GRAPHICS

NAME:

FILE NO.:   SECTION:

GRADE

DIMENSIONING

DIM~4

DIMENSION THE OBJECTS. EACH GRID = 3mm OR 1/8". NOTE THE FINISH MARKS.

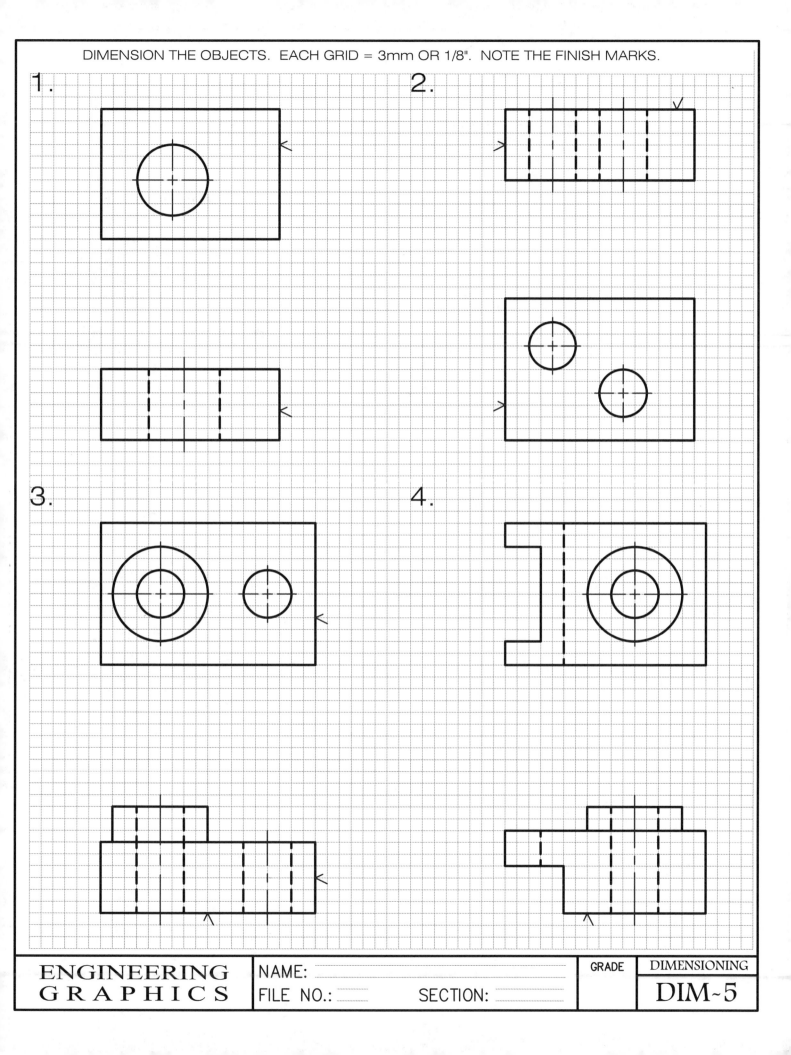

1.

2.

3.

4.

ENGINEERING GRAPHICS	NAME: _____ FILE NO.: _____ SECTION: _____	GRADE	DIMENSIONING
			DIM~5

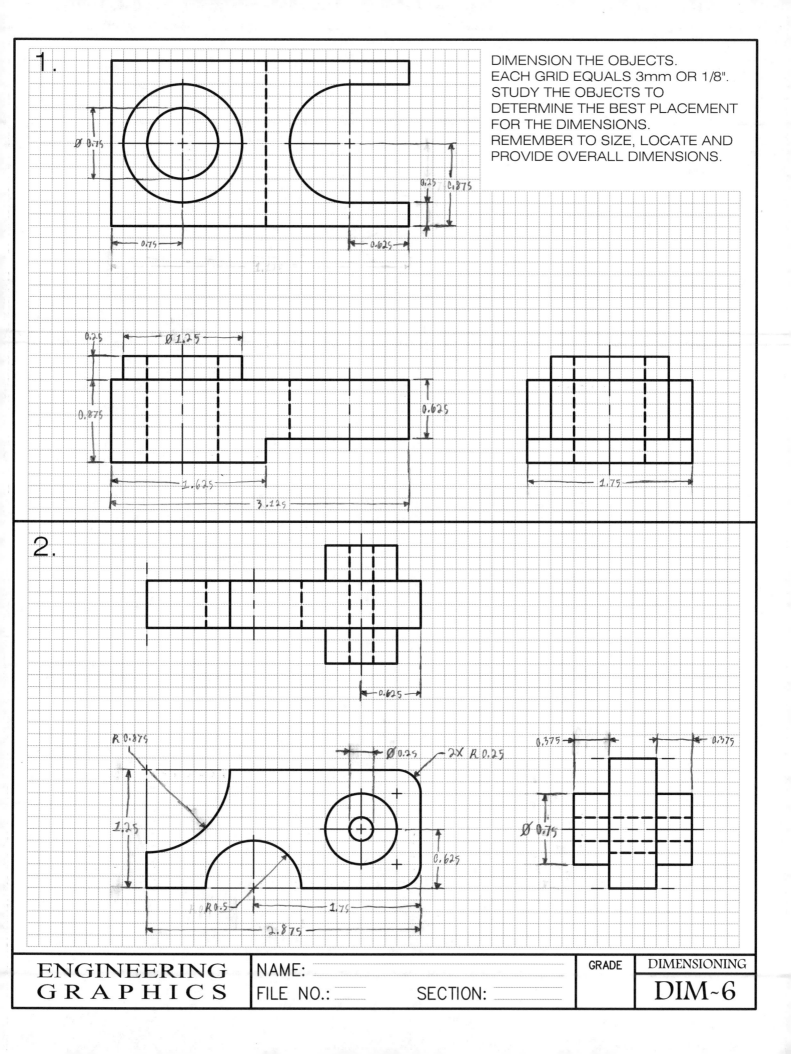

1.

DIMENSION THE OBJECTS.
EACH GRID EQUALS 3mm OR 1/8".
STUDY THE OBJECTS TO
DETERMINE THE BEST PLACEMENT
FOR THE DIMENSIONS.
REMEMBER TO SIZE, LOCATE AND
PROVIDE OVERALL DIMENSIONS.

Ø 0.75

0.25 Ø.875

0.75   0.625

0.25   Ø 1.25

0.875   0.625

2.625   1.75

3.125

2.

0.625

R 0.875   Ø 0.25   2X R 0.25   0.375   0.375

1.25   Ø 0.75

0.625

R 0.5   1.75

2.875

ENGINEERING
GRAPHICS
NAME:
FILE NO.:   SECTION:
GRADE   DIMENSIONING
DIM~6

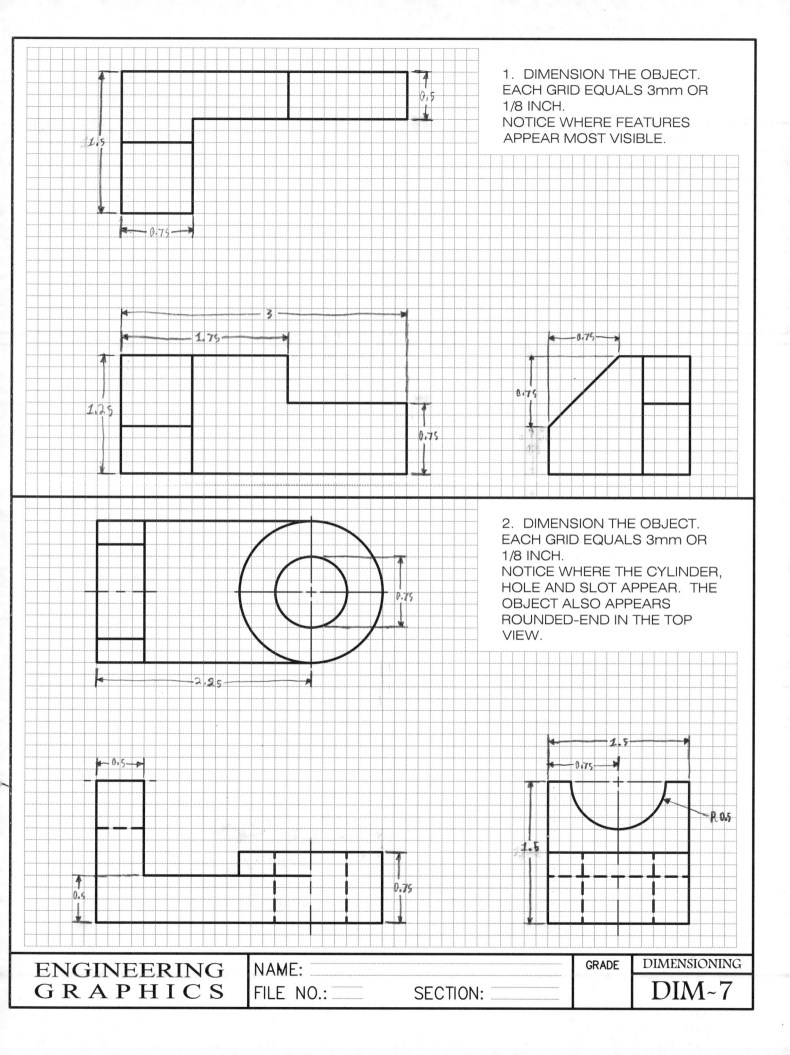

1. DIMENSION THE OBJECT.
EACH GRID EQUALS 3mm OR
1/8 INCH.
NOTICE WHERE FEATURES
APPEAR MOST VISIBLE.

2. DIMENSION THE OBJECT.
EACH GRID EQUALS 3mm OR
1/8 INCH.
NOTICE WHERE THE CYLINDER,
HOLE AND SLOT APPEAR. THE
OBJECT ALSO APPEARS
ROUNDED-END IN THE TOP
VIEW.

ENGINEERING
GRAPHICS

NAME:
FILE NO.: SECTION:

GRADE | DIMENSIONING

DIM~7

Dimension the objects. Each grid equals 3mm OR 1/8". Both objects are rounded-end.

1.

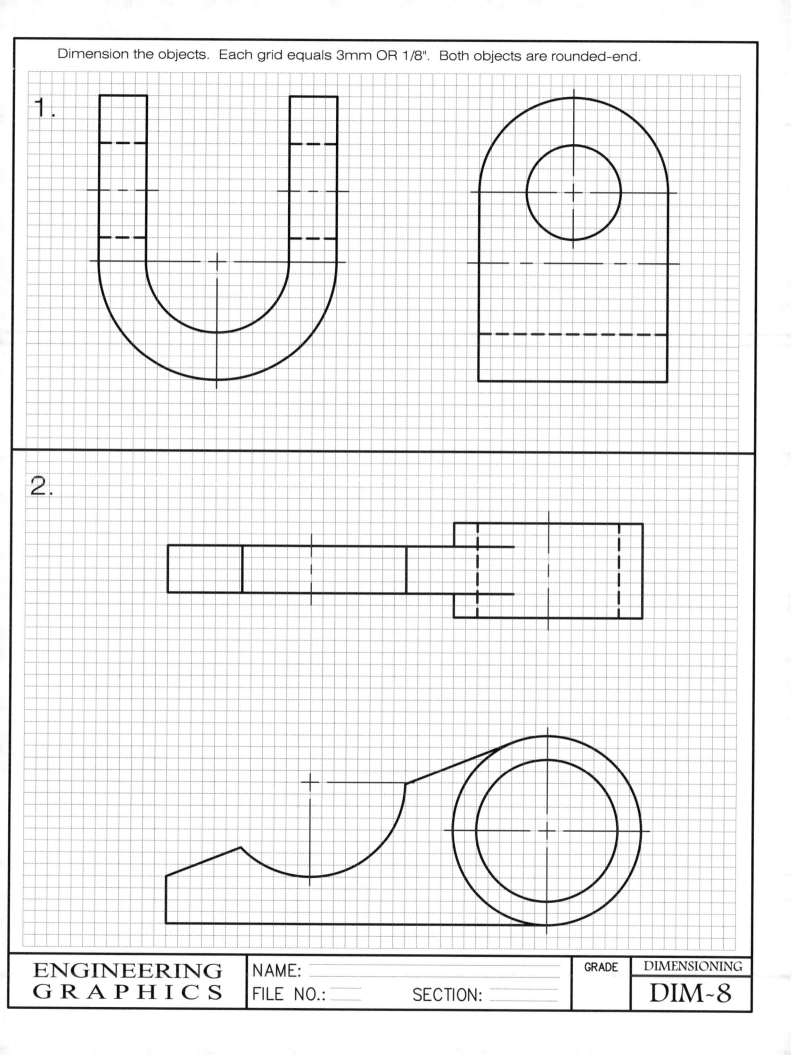

2.

ENGINEERING
GRAPHICS

NAME:

FILE NO.:   SECTION:

GRADE

DIMENSIONING

DIM~8

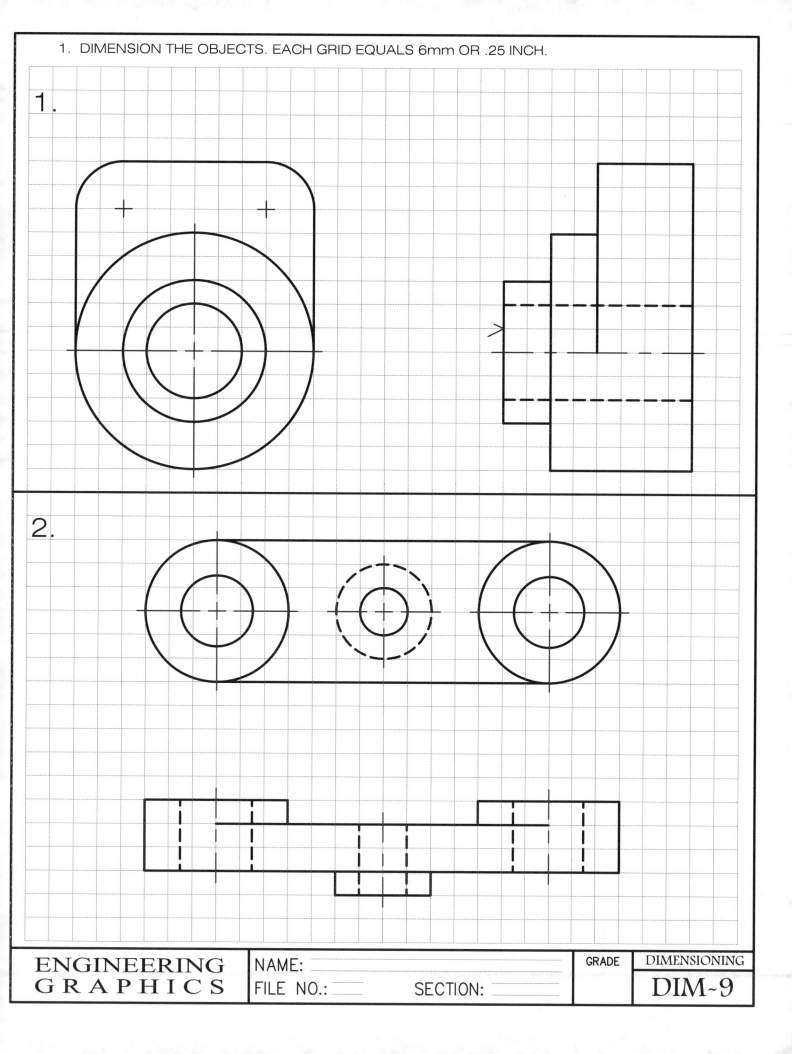

1. DIMENSION THE OBJECTS. EACH GRID EQUALS 6mm OR .25 INCH.

1.

2.

ENGINEERING GRAPHICS

NAME: _____

FILE NO.: _____ SECTION: _____

GRADE

DIMENSIONING

DIM~9

DIMENSION THE OBJECTS. EACH GRID = 3mm OR 1/8".

1.

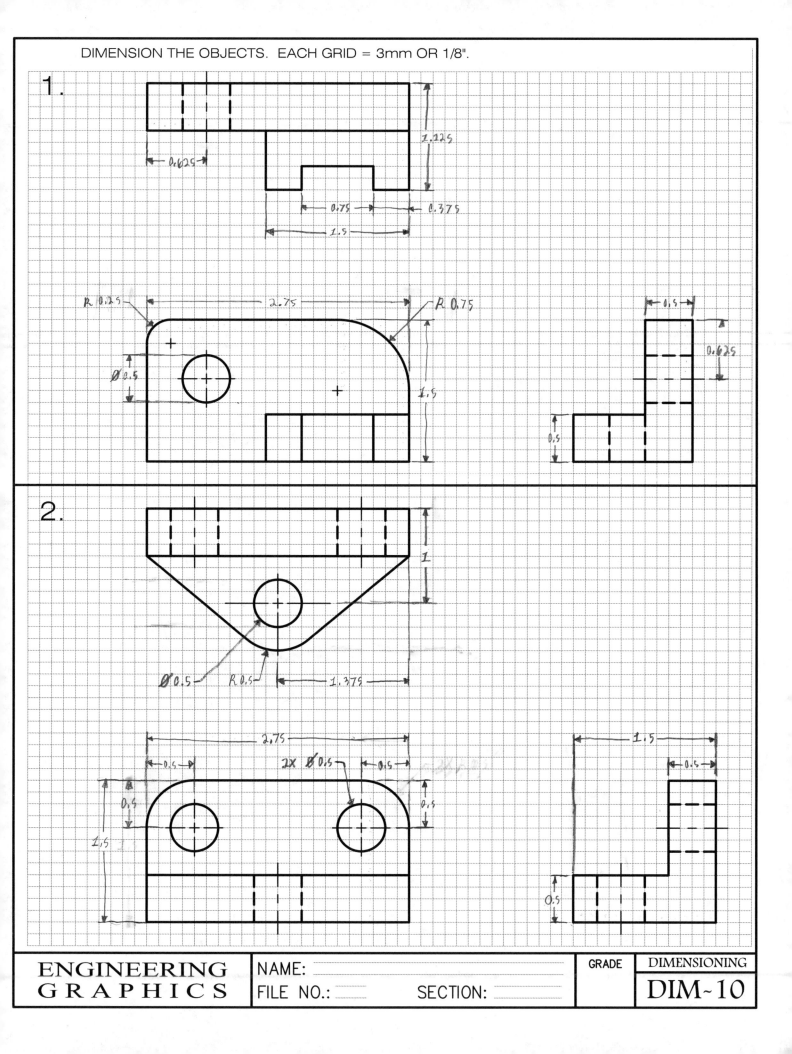

0.625

1.125

0.75   0.375

1.5

R 0.25   2.75   R 0.75

Ø 0.5   1.5

0.5

0.625

0.5

2.

1

Ø 0.5   R 0.5   1.375

2.75   1.5

0.5   2X Ø 0.5   0.5   0.5

0.5

1.5   0.5

0.5

ENGINEERING
GRAPHICS

NAME:
FILE NO.:   SECTION:

GRADE   DIMENSIONING

DIM~10

Dimension the given object. Study
the features to determine how to
apply the basic rules of dimensioning
with reference to rounded-end
objects, multiple holes, most visible
views, radiused corners and finished
surfaces. Use a scale of 1 grid equals
6mm or .125".

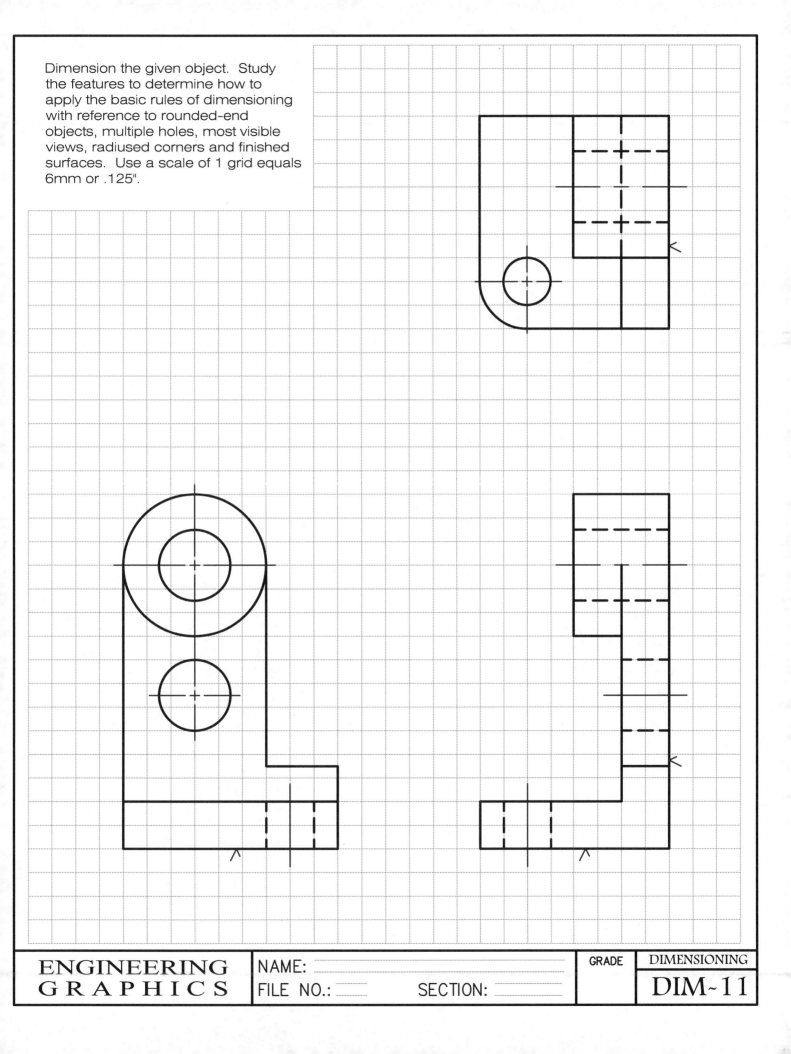

ENGINEERING
GRAPHICS

NAME:

FILE NO.:   SECTION:

GRADE   DIMENSIONING

DIM~11

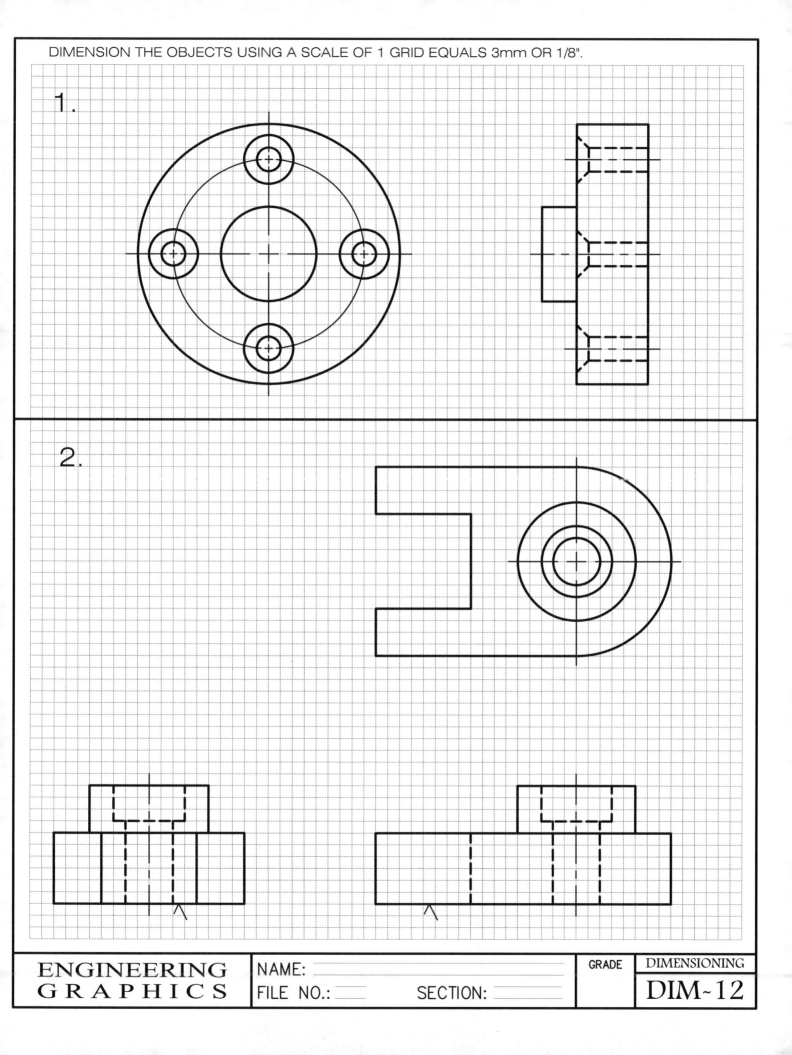

DIMENSION THE OBJECTS USING A SCALE OF 1 GRID EQUALS 3mm OR 1/8".

1.

2.

ENGINEERING
GRAPHICS

NAME:

FILE NO.: SECTION:

GRADE

DIMENSIONING

DIM~12

DIMENSION THE OBJECTS. EACH GRID EQUALS 6mm OR .25 INCH.

1.

NOTE:
THE PLATE IS 9MM THICK.

2.

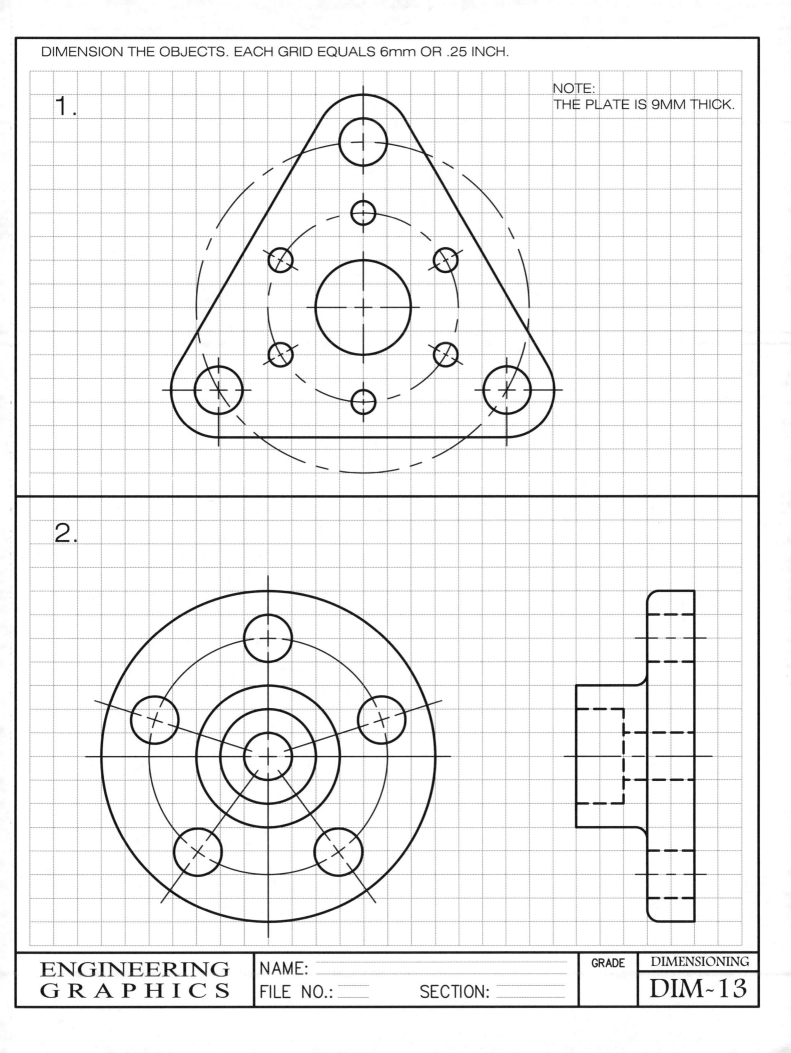

ENGINEERING
GRAPHICS

NAME:

FILE NO.:

SECTION:

GRADE

DIMENSIONING

DIM~13

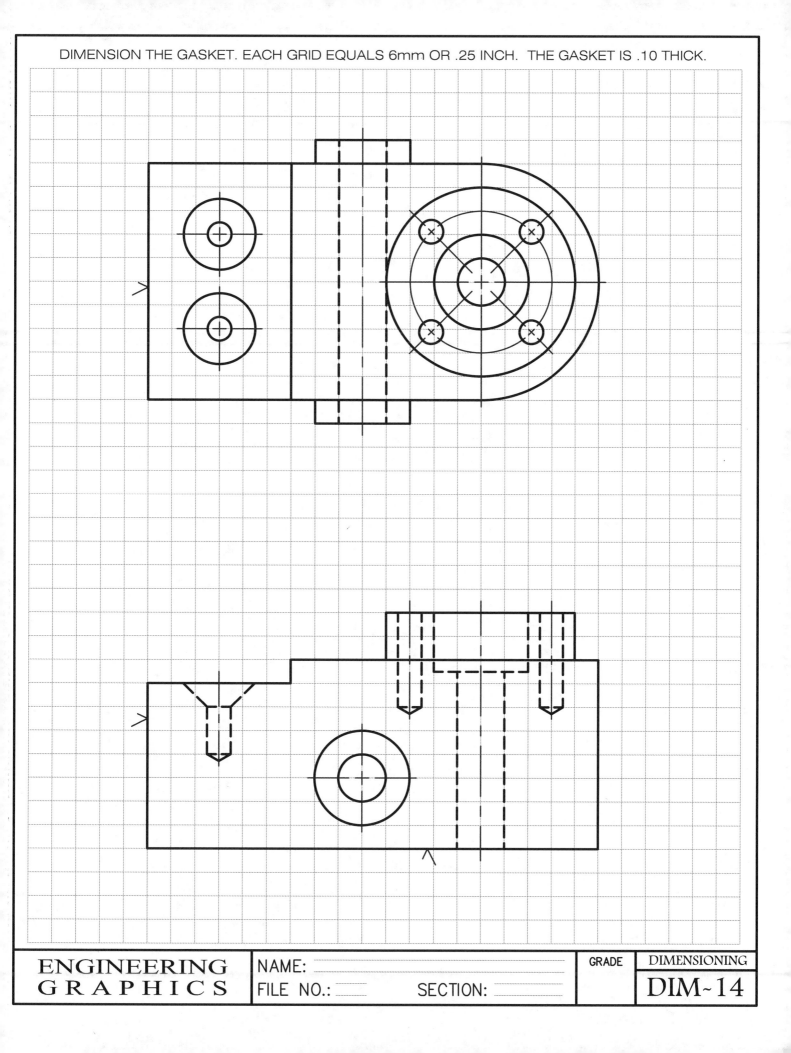

DIMENSION THE GASKET. EACH GRID EQUALS 6mm OR .25 INCH. THE GASKET IS .10 THICK.

ENGINEERING
GRAPHICS

NAME:

FILE NO.:   SECTION:

GRADE

DIMENSIONING

DIM~14

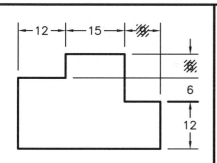

Omit one dimension in a series.

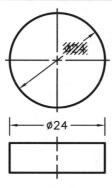

Dimension cylinders where they appear rectangular.

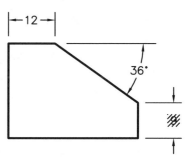

For an angled surface, use two coordinates or a coordinate and angle.

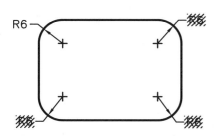

Locate the cylinder where it appears circular. Do not dimension the outer edge.

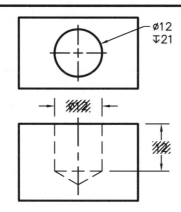

Provide the diameter and depth for the hole in the note.

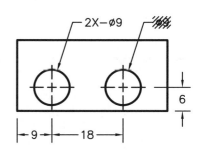

Indicate the number of identical holes with the number followed by "X" for "times". Locate the holes where they appear circular

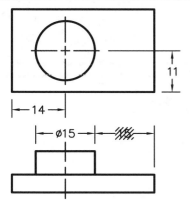

Give only one radial dimension if they are all identical.

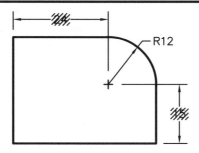

The radius is self-locating.

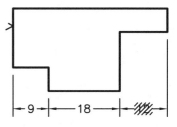

Always dimension from a finished surface if one is indicated.

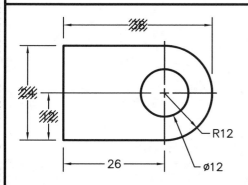

Omit overall dimensions for rounded-end objects. Omit dimensions already indicated by the radius.

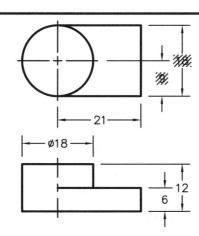

Omit measurements that can be easily calculated from given diameters.

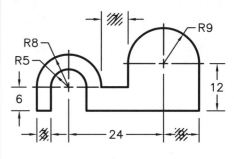

Omit dimensions that are "left-over" as a result of given dimensions.

DIMENSIONING

DIM~15

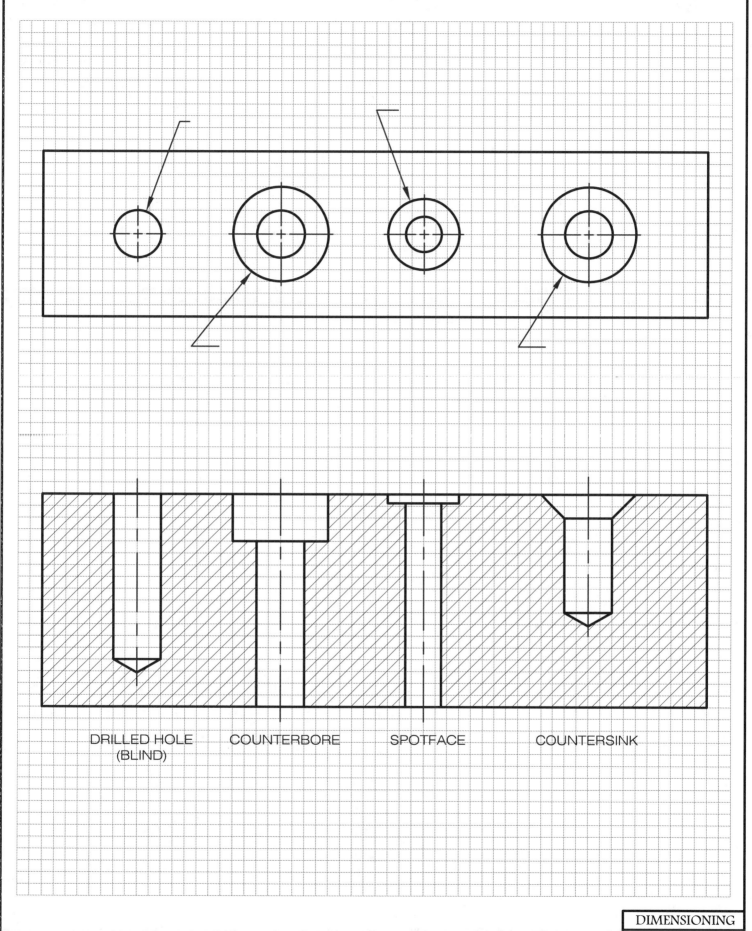

DRILLED HOLE
(BLIND)   COUNTERBORE   SPOTFACE   COUNTERSINK

DIMENSIONING
DIM~16